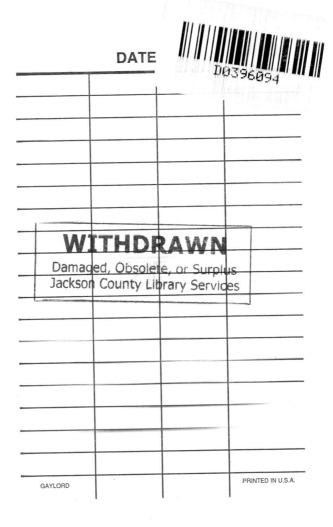
This book made possible by a
contribution to the
Jackson County Library Foundation's

Buy-A-Shelf campaign

from a reader like you!

A Touchstone Book   Published by Simon & Schuster

SKIES of FURY

Weather Weirdness Around the World

Patricia Barnes-Svarney & Thomas E. Svarney

TOUCHSTONE
Rockefeller Center
1230 Avenue of the Americas
New York, NY 10020

Designed by Gabriel Levine

Manufactured in the United States of America

10 9 8 7 6 5 4 3 2 1

Library of Congress Cataloging-in-Publication Data

Barnes-Svarney, Patricia L.
    Skies of fury : weather weirdness around the world / Patricia
Barnes-Svarney & Thomas E. Svarney.
        p.   cm.
    "A Fireside book."
    Includes index.
    1. Weather—Popular works.   I. Svarney, Thomas E.   II. Title.
QC981.2.B35   1999
551.5—dc21   99-29956
                                                                CIP

ISBN 0-684-85000-1

To Barb and Gary Snyder, best friends—
for years of love, laughter, and Thousand
Islands sunsets. . . .

Acknowledgments

Many people have been kind enough to help us through this book. We would like to thank our editor Matt Walker, for his talent and patience in helping us with the manuscript. Thanks to editor Sarah Pinckney Whitmire for her encouragement and help; and to publisher Trish Todd for seeing the potential of this project. And of course, as always, thanks to our agent, Agnes Birnbaum, for her guidance all these years.

The authors would also like to thank the many weather organizations that helped us along the way, including the kind people at the National Center for Atmospheric Research, in Boulder, Colorado; the National Oceanic and Atmospheric Administration, especially in Asheville, North Carolina; and the Mount Washington Observatory. Thanks, too, to Joe Sobel of Accuweather for years of great weather coverage on our local radio station, WNBF (and a special thanks to Bill Parker), in Binghamton, New York (and many other

ACKNOWLEDGMENTS

places around the country). And, finally, thanks to all the amateur and professional weather people who, for more than a century, have diligently recorded, watched, and tried to explain the topsy-turvy world of weather.

Contents

Preface

As scientists, we're exposed to the weather more than usual as we travel in the field. Most of the time, we can do our work on either a cloudy or sunny day. Or we can endure a warm rain as we explore. But an ice storm can also mean a canceled flight, or a string of hot days can cause the trains to be late.

Of course, we're not the only ones. *Everyone* is affected by weather. For most people, weather, the local atmospheric changes we experience every day, is usually viewed with joy or disgust. We praise weather when it treats us to a sunny and warm day for a family picnic, and curse it when sleet falls on the day we chose to travel. Weather is also a matter of debate, some preferring hot-weather beachcombing to cold-weather skiing. We tend to criticize the weather forecasters for not predicting or controlling the weather, as if they really had a say in the matter. And of course, many of us are addicted to the weather channels, on television and the Internet, places that give us on-the-spot pictures of what weather is coming our way, eliminating the surprise of the "coming-out-of-nowhere" weather events our ancestors had to endure.

But even with all our warnings and colorful weather maps, the weather continues to awe and amaze us with its power and weirdness. One reason arises from the very nature of weather: Unlike other physical events on Earth, such as the lava flow from a volcano or the shaking of an earthquake, we can't really "touch" weather. We can't see or touch the air rising to build a thunderstorm, and we have no way of watching how positive and negative charges meet to form a lightning bolt. The atmosphere becomes the playground for nature's magicians. And we are the passive observers of their demonstrations—from calming rainbows to skies of fury.

Whether you are traveling the Rocky Mountains of the United States, the deserts of Australia, the Caribbean islands, or the windblown ice sheets of the South Pole, there are always weather sights particular to a region. This book starts out by explaining the better-known weather events and some of the weirder facts about these common phenomena. It then swings its way through the many "sky shows" that develop in other parts of the world, revealing the tricks of light and optical illusions of these weather events. We explain some of the stranger weather phenomena and patterns that often pop up in geographically diverse regions, and even give you a few Internet weather sites to explore for more information.*

Weather is important to us all. It has been for thousands of years, and it will continue to be so in the twenty-first century and beyond. Intuitively, we all know that without weather and its connection to the Earth's natural cycles, such as precipitation, evaporation, and condensation, life would cease to exist. Maybe our need to know the weather and explain its many forms is based on one fundamental idea: The weather we see truly represents not only the events driving the Earth and its cycles, but life as we know it.

---

*Please note: Many Internet sites—and addresses—change in a short period of time. If you experience any difficulty trying to link to a site, try searching for a key weather word in a search engine. We apologize for any inconvenience if you obtain an out-of-order link site.

1

Familiar (and Unfamiliar) Weather Everywhere

Our planet's orbit is just close enough to the Sun to give us the warmth we need for life to survive, and to keep the world's water in three necessary states: gas, liquid, and solid. Conversely, we're not close enough to our star to burn off the oceans and atmospheric gases into the voids of space, as happened on the planet Mercury; or close enough to produce a runaway greenhouse effect like that on the planet Venus. As living organisms, we are, on this third rock from the Sun, in the right place at the right time. But there is an aftereffect of being at this point in orbit around our star: weather, good and bad, with all its associated phenomena.

Earth's Weather Engine: Weather or Not?

Think of yourself as an apprentice weather magician. Your first task is to make the Earth a sphere with no continental land-masses. Next, take water molecules and create conditions of 100 percent liquid, filling the deep oceans with water. At the ocean surface, throw in a planetary atmosphere in which gases such as nitrogen and oxygen increase and water vapor diminishes to less than 1 percent. Finally, decrease the water vapor with increasing height, until it essentially reaches zero at the vacuum of outer space.

Sound simple? It is, but this is not the world in which we live. The perfect world you just put together would be predictable and somewhat dull (no offense — it *was* your first try), and extremely unlike the real Earth. Next, you'll need a few important ingredients in order to receive your weather magician diploma: the heat from the Sun, the planet's rotation and tilt, and a few large and small landmasses.

Why do we point fingers at the Sun? This glowing orb of light is the major component in the planet's weather engine as it splashes energy in the form of heat toward our planet. The effect of this energy on the Earth is somewhat similar to heating a pan of water over a hot stove: Heat warms the water at the pan's bottom, which then rises to the top, while the cooler water on top sinks to the bottom. This continual movement of water in the pan, caused by the action of the external heat, is known as *convection*.

The Earth's atmospheric convection starts as the Sun's heat plunges through the atmosphere, falling most directly on the equator. Here, the incoming heat exceeds the outgoing heat in a thick band around the belly of the Earth. The air warms and rises, then cools and condenses, producing clouds and heavy rainfall typical of the equatorial regions. As the air rises higher toward a layer of the atmosphere called the *tropopause* — about 7 miles (11 kilometers) above the Earth's surface — the air is deflected poleward. At about 30 degrees latitude, this air cools and begins to sink again. And as the air descends, higher pressures toward the surface wring out the air's moisture like a sponge, creating dry subtropical deserts underneath, such as Africa's Kalahari and Sahara.

As the dry, sinking air is sucked back toward the equator, the *Hadley cells*, the first convective loops, are complete. This is caused by a well-known axiom in science: Nature abhors a vacuum. In this case, the surface air rising at the equator creates the "vacuum," and dry air rushes in. In turn, this rushing produces winds, the northeast and southeast *trade winds* in the Northern and Southern Hemispheres, respectively.

The next loops are the *Ferrel cells*, created as some of the descending air at about 30 degrees latitude continues toward the respective poles; the air rises as it clashes against the cold polar air. The "vacuum" caused by this loop forms the mid-latitude winds called the *westerlies*. In the last northern and southern loops, cold air sinks at the poles and rises as it hits the Ferrel cells. This loop, called the *polar cell*, creates the winds called the *polar easterlies*. When the warmer, moist air from the middle latitudes clashes with the cold, dry air from the poles, a battle often ensues, leading to the familiar major rain or snowstorms in the mid-latitudes.

There's more to this atmospheric engine. Similar to a chicken on a barbecue spit, the Earth rotates on an axis that tilts 23.5 degrees to the plane of our orbit around the Sun. This causes our star's direct rays to fall on certain specific portions of the planet during the year—the reason for our four seasons—and results in differential heating that mixes our atmosphere even more. Winters at the poles are good examples of the effects: The differences between the temperatures at the poles and equator become more pronounced, with strong winter winds and storms pushing the weather engine even faster.

The Earth's rotation, in addition to giving us night and day, adds a horizontal component by tugging and twisting global wind patterns—the westerlies, easterlies, and trade winds—into a more east or west direction, depending on the hemisphere. If you could actually see air movement from high above the Earth, in the Northern Hemisphere the winds (and water currents) would deflect to the right (east), and in the Southern Hemisphere, to the left (west). Thus, winds flowing toward the equator from both hemispheres move in an east-to-west direction; winds that flow toward the poles from the Tropics of Cancer and Capricorn move in a west-to-east direction; and finally, the pole winds flow from east to west. These winds shift a little to the north and south seasonally, as the Sun moves back and forth across the equator.

But we aren't finished yet. Now we must dot the planet with continents and larger islands. And on these landmasses, pitch in diverse features, such as snowfields, coastal plains, tropical forests, high mountains, or dry deserts. All of these impact the amount of heating or cooling of the atmosphere above them, which in turn influences the movement of the air masses and the amount of water vapor above a region.

You are now a journeyman weather magician. You know the *how* behind the giant planetary circulation. Simply put, the complex engine is now started, the system is running, and so is our weather. And by tweaking the rainfall, increasing the winds, or making a hundred other iterations, you'll soon see why strange weather phenomena pop up all over the world.

• For an overall weather link, try WeatherNet:

http://cirrus.sprl.umich.edu/wxnet/

• Cycling Water •

Weather systems around the world are a direct result of the water cycle, the endless movement of water from the atmosphere to oceans and land, and back. When it rains, water vapor condenses into a liquid. The liquid falls to the surface and is either carried as runoff or into the soil, most often to recharge the groundwater; other droplets land on plants that need the rain for growth and exchange of oxygen and carbon dioxide; and still other drips find their way back into the atmosphere. Such seemingly simple things as precipitation, humidity, and evaporation all become necessities for life as we know it.

Amazingly, the amount of water the cycle has to play with is minute: A whopping 97 percent of all the Earth's water sits in the oceans, 2 percent is frozen in glaciers and ice sheets, and less than 1 percent flows in streams, lakes, and as groundwater. Only 0.001 percent of the planet's water is in the atmosphere at any one time, and yet it serves as the key to the movement of our atmosphere and the world's weather.

Humidity: It's the Humidity

Though some scientists scoff at the idea, people with rheumatism or arthritis claim their joints swell and ache when a humid air mass is approaching. Likewise, asthma attacks can reach epidemic numbers during times of high *humidity*, but for an indirect reason: In the summer of 1983, for example, a large rise in hospital admissions for asthma attacks occurred in England after the humidity from a storm released tiny fungal spores, triggering the attacks. Despite its frequently complicated effects, humidity—the sticky, oppressive air that makes most of us stay in front of a fan or air conditioner in the summer—is simply the amount of water molecules in the air.

Interestingly enough, air with more water vapor (or a higher humidity) weighs less than dry air at the same temperature and pressure, a fact known even to Sir Isaac Newton, who mentioned it in his book *Optics* in 1704. Our perception that humid air feels oppressive, or "heavier," has to do with less oxygen for breathing and our inability to get rid of sweat by evaporation, not the air's actual weight.

Meteorologists use two ways of describing humidity. The first is *relative humidity* (RH), or the amount of water vapor present in the air relative to the amount the air could actually hold, given its current temperature. The resulting ratio multiplied by 100 gives the relative humidity as a percent. In any geographical area, temperature can cause the RH to bounce up and down the scale. Under certain conditions, a lower temperature can mean 100 percent relative humidity; boosting the temperature 20 degrees can decrease the RH to 50 percent.

And if you're in the mood for an RH of *over* 100 percent, try living in a cloud.

Another way to report humidity is by the *dew point temperature*, or the temperature at which the moisture in the air would condense (essentially turn into a cloud) if it cooled gradually. The higher the dew point temperature, the greater the amount of moisture in the air. If the air temperature falls to this reading or below, the relative humidity will reach 100 percent and water vapor will start to precipitate out as a liquid. This is why weather reporters often mention the dew point temperature and the potential for rain in the same breath; when the atmosphere cools, the extra water vapor condenses into liquid and falls as precipitation.

Humans feel somewhat uncomfortable with dew point temperatures in the mid-60s. But dew points higher than the 70s are dangerous—something people who live in the tropics or even the hotter and humid regions of the world all know. High dew point temperatures mean a reduction in surrounding oxygen, and people with upper respiratory problems suffer. The body tries to regulate its internal temperature through sweating, but the moisture has a hard time evaporating into the water-laden air. Any strenuous activities become impossible and dangerous, potentially resulting in heat exhaustion or heat stroke.

If you find yourself wanting to go on a summertime jog, you may want to avoid the oppressive tropical jungles of the world—in particular, parts of the Ethiopian coastline along the Red Sea. There, the region's dew point reaches an average of 84 degrees in June. To compare, in the United States, a dew point temperature of 70 degrees is oppressive and sticky. During the rare readings that top a dew point temperature of

80 degrees, weather forecasters tell people to turn on the air-conditioning and stay inside.

Temperature: It's Not the Heat

Temperature is probably the one thing we're aware of no matter where we stand, walk, sit, or lie. Comfortable indoor or outdoor temperature, like your favorite color, is a matter of personal preference. The comfort or discomfort when experiencing different temperatures in our surroundings is based on our own body—affected by its metabolism and even the climate in which we were born and raised.

The definition of *temperature* depends on your frame of reference. Physicists and chemists refer to temperature as the standardized measurement of air molecule movement—not a subjective feeling of comfort or discomfort. The more energy or heat, the faster the movement of the molecules and the higher the corresponding temperature. And much of the time, scientists deal with temperatures beyond our ken. For instance, physicists often discuss *absolute zero*, the temperature at which there is no movement of air molecules, at about −459.67°F (−273.16°C). Of course, this is far below a cold winter's day even in the Antarctic.

We're all more familiar with the term *surface temperature*, the temperature at a certain place and time on the surface of the Earth. Logically, surface temperature differs from point to point on the Earth, changing with altitude, the amount of water vapor present, and the local wind speed. For example, the temperature decreases an average of about 1 degree

Fahrenheit for every 100 feet (30 meters) of altitude, extending from the Earth's surface to about 45,000 feet (13,716 meters). This is also why it may be warmer at your house in comparison to your local weather station's reading, since many such stations are located on hilltops.

There is one way that meteorologists *can* relate to us: by using the *heat index*. By looking at the combined effects of temperature and relative humidity, meteorologists chart the apparent temperature, or the temperature your body observes. When the relative humidity is high, the body is almost smothered by the blanket of its own perspiration, since the body cannot cool off by evaporation of sweat. It seems warmer to the person, and thus the heat index temperature is higher than the actual temperature. A lower humidity would allow sweat to evaporate and cool down the body; thus the heat index temperature would be lower than the actual temperature. In the hot summer months, most of us can relate to the higher heat indexes. If the relative humidity is 100 percent and the temperature 85°F (29.4°C), the heat index is 108°F (42.2°C)—dangerous for almost any activity outdoors. At 80°F (26.7°C) and only 50 percent relative humidity, the heat index is more reasonable: 81°F (27.2°C).

Speaking of temperature, have you ever wondered why it seems hotter in some cities than in the surrounding countryside? Many times, this is due to a recently discovered phenomenon called *heat islands*, areas of localized high temperatures, typically 1 degree higher around smaller cities and up to 4 degrees higher around larger cities. The culprits are humans: Most of the vegetation around the larger cities has been replaced with asphalt, concrete, and buildings, which absorb heat during the day and radiate it at night. The local weather in St. Louis, Missouri, changed dramatically after construction eliminated a great deal of vegetation; there were more thunderstorms, hail, and rainfall. Other cities that kept their natural vegetative covering, such as Columbia, Maryland, experience only a small increase in local temperature and no corresponding change in weather patterns.

• Hot and Cold Records •

We all know prolonged heat waves can make us irritable, and one study shows just how irritable we can get. The murder rate increased by approximately 75 percent in New York City during a thirty-two-day heat wave in 1988. The highest temperature ever recorded on Earth measured 136.4°F (58°C) and was recorded at Al' Azizyah, Libya, in the Sahara Desert. Fortunately, the population there is relatively sparse.

According to British and American records, the two hottest years on record were 1990 and 1991; in the Northern Hemisphere, temperatures in the years 1990, 1995, 1997, and 1998 exceeded all others since 1400. The hottest month in the Northern Hemisphere so far was July 1988. The warmest two winter months on record in the United States came in January and February 1998. The 1980s was the world's warmest decade of the twentieth century, though the 1990s will no doubt pull out in front of the pack.

The coldest prize goes to the Vostok Station, Antarctica, in 1983, reaching the planet's lowest temperature on record: −128.6°F (−89°C). In 1954, the United States recorded its lowest temperature at Rogers Pass, Montana: −70°F (−57°C). And Antarctica also holds another temperature record: the lowest average temperature per year, measuring −71.7°F (−57.6°C).

• Counting Chirps •

Outdoors without a thermometer, some people are sure they can tell the local temperature in the summer by counting the number of times a cricket chirps. Crickets apparently chirp 72 times per minute at 60°F (15.6°C). For every additional 4 chirps, add a degree to the 60°F, and for every 4 less, subtract a degree. Or just count the number of chirps in 14 seconds, then add 40. We've tried both these methods and they seem to work (relatively speaking). The only problem is, sometimes the chirping comes from seemingly dozens of crickets—making it impossible to pick out a single cricket's voice amid a chorus of chirps.

Air Pressure: Contents Under Pressure

Many of us are under pressure these days to fulfill some aspect of our lives, whether it is family, work, or play. But what most people don't realize is that we are, and always will be, under another type of pressure—air pressure. Billions of molecules constantly move in all directions, bouncing off of us and other objects around us. Gravity pulls the air molecules tighter as you get toward the Earth's surface, increasing the pressure; conversely, as you climb in altitude, the number of molecules, thus the pressure, decreases.

Air pressure also changes as certain air masses, or pressure systems, of different temperatures and humidities pass through a region, usually with enough power to influence a change in local weather. Where lighter, humid air rises, a *low-pressure system* develops, spiraling upward in a counterclockwise direction in the Northern Hemisphere. Dry air cools and drops in

25

high-pressure systems, spiraling down toward the ground in a clockwise motion in the Northern Hemisphere, compressing and warming the air as it descends. These spiraling systems spin in the opposite direction in the Southern Hemisphere: counterclockwise and clockwise for the high and low pressures, respectively.

High-pressure systems usually bring in dry, sunny skies. But the low-pressure systems can be formidable. Between March 6 and 7, 1932, a deep low-pressure system moved up the East Coast, one of the lowest low-pressure systems ever recorded. The system set many records at that time, measured in the old method: barometric readings of 28.78 inches in Charleston, South Carolina; 28.54 inches in Philadelphia, Pennsylvania; and 28.45 inches in Boston, Massachusetts. Compare these readings to an average pressure of about 30 inches on a bright, sunny day.

• For some of the best Internet daily weather sites to keep track of
high-pressure and low-pressure systems, just link to
The Weather Channel, Accuweather, USAToday weather,
CNN weather, or any other large broadcasting group.

High-pressure and low-pressure systems all spin in certain directions, depending on the hemisphere. These two low-pressure systems in the Bering Sea, between Alaska and Asia, spin in a counterclockwise direction. (courtesy of NASA, photographed by the space shuttle *Discovery* in 1994)

Fronts: It's What's Up Front That Counts

Like two fighters in a boxing match, collisional boundaries between two air masses, especially in the temperate zones, bring striking weather phenomena to our skies. Called *warm fronts* and *cold fronts*—named for the battle fronts of World War I—they can carry some of the nastiest weather on the planet in the middle and northern latitudes.

Take, for example, the squall line that developed well in advance of an eastward-moving cold front on the night of September 3, 1925, as it collided with a warm front. The midwestern storm doomed the USS *Shenandoah*, a lighter-than-air ship fashioned after the German zeppelin. Aloft and unaware, no one knew the front was approaching until thunder and lightning appeared, the storm developing rapidly as a cold northwest current of air overran a warmer southwest current. The resulting mixing of warm and cold air wrenched away two sections of the airship held together by the control cables along the keel. The control car broke from the frame, tumbling to the ground and killing the eight men onboard. Eventually, the turbulence caught the nose of the airship, subjecting the hull to unequal forces and fracturing the ship into three parts. The engine gondolas let loose, killing four of the mechanics. The tail section fell against a wooded hillside, and eighteen people in the tail gasbag scrambled to safety. The seven men in the remaining balloon gasbag hung on as the winds died down, permitting a landing.

Warm and cold fronts meet, creating "weather wedgies," as the air masses slide under or over (overrunning or overriding) each other. For example, when the tip of an advancing cold front wedges under a warm front, pushing the humid air up quickly, the dramatic rise of the warmer air gives birth to immense thunderstorms. In the Northern Hemisphere, gusty winds are often associated with cold fronts, usually coming from the south or west directions (in the Southern Hemisphere, from the north and west), while temperatures either remain steady or rise.

Another wedgie forms as air from an unusually slow-moving warm mass runs up and over a stationary cold air mass at the surface. As the warmer, lighter, less-dense air advances along the cold front, it gradually rises higher and higher, cooling until the water vapor condenses, forming clouds and eventual precipitation. If the air beneath is at or below the freezing point, such as in the winter, precipitation falls as snow or ice; if it is summer, precipitation falls as rain. As this type of front rolls through in the Northern Hemisphere, the wind shifts around from the east to south (east to north in the Southern Hemisphere), with the temperature rising.

There are also more-complex fronts. An *occluded front* forms as a cold front overtakes a warm front. A *stationary front*, or quasi-stationary front, is just what it sounds like—a front traveling at near-static speeds. Other, more specific fronts, occur too. *Upper fronts* are present in the upper atmosphere but do not extend to the ground; *anafronts* occur when warm fronts reach into the high altitudes and *katafronts* when they descend (usually as cold fronts) from high altitudes. And there

Towering cumulonimbus calvus clouds forewarn an approaching cold front, as seen from a glacial ridge in the Finger Lakes region of New York. The mushroom shapes indicate vigorous updrafts pushing the clouds upward. These clouds always produce some type of precipitation.

are strange fronts that are not well studied, such as the *intertropical fronts* (equatorial or tropical fronts) that appear to exist in a band near the equator, separating the air masses of the Northern and Southern Hemispheres.

• Understanding a Television Weather Map •

Not every television weather map is the same, but there are some general characteristics that can help you better understand the weather coming your way. The centers of high-pressure systems are represented by the large red *H* on your screen, spinning clockwise in the Northern Hemisphere; the counterclockwise low-pressure systems are represented by the blue *L*. The blue lines represent a cold front, studded with triangles pointing in the direction of movement. The red lines spotted with semicircles represent a warm front. A stationary front is represented by alternating blue triangles and red semicircles on the respective sides of a red-and-blue line. A dashed purple line indicates an upper-air disturbance, usually in association with a low-pressure system. Faint white lines are *isobars* connecting points of similar barometric pressures. Surface wind directions run almost parallel to the isobars, and the closer the isobars, the stronger the wind speeds, owing to greater pressure differences.

Even if you know how to read a weather map, don't feel bad if you still can't predict the weather. After all, weather reports can vary among forecasters, the way several doctors can make slightly different diagnoses from the same lab report. Amazing as it may seem—and contrary to what most people think—it's estimated that weather forecasters are correct 85 percent of the time.

A frontal boundary of a storm is seen as a straight line of clouds, an image taken over Argentina, on space shuttle mission STS-43. (courtesy of NASA)

• When Backdoor Fronts Come Knocking •

Not all cold fronts travel from west to east in the United States. One strange type of front, called *backdoor cold front*, originates in the northwestern Atlantic Ocean and moves to the northeast. These are usually associated with high-pressure systems circling clockwise just off the coast and manage to push in cooler marine air toward the land. Sea breezes (which flow from the cooler ocean to the warm land during the day) exacerbate the effect, pushing the warmer, and thus lighter, air back inland—adding to the cooldown of the coast. These weather patterns usually occur during the early spring, when the temperature differences between the cooler sea and warming land are greatest.

Backdoor fronts are often the reason why people curse the

return of winter: After several warm spring weeks the air turns cold again, leading everyone in the northeast to believe warmer weather will never come. But backdoor cold fronts can open up during the summer months too, pushing hotter air inland and cooling down the coast, often with temperatures dropping a dozen degrees or more.

Jet Streams: Jet-Setting Around the World

Jet streams are upper-level winds formed at the boundaries between major air masses, steering the giant weather patterns around the planet. They were postulated in the early 1940s by scientists, but none was actively encountered until the end of 1944, when high-cruising American B-29 aircraft flew over Japan during World War II. The pilots found the west-to-east moving rivers of fast air as over 100 planes were sent to bomb the industrial sites near Tokyo. As they approached the city at a height of about 6 miles (10 kilometers), something pushed the planes at speeds of up to 150 miles (241 kilometers) per hour.

Where the large, different-temperature air masses meet high above the Earth's surface, a great deal of energy builds up in a wavy transition zone. And just like a young child with stored-up energy, the atmosphere tries to release it. In this case, the release is in the form of very fast moving west-to-east winds at high altitudes. These jet stream winds weave their way around the planet, contributing to the overall atmospheric heat exchange, easily transferring warm tropical air toward the poles and cold polar air toward the equator. Overall, there are four major jet streams—two each in the Northern and Southern Hemispheres—all of which weaken in the hemispheres' respective summers and strengthen in the winters.

A storm's intensity can be exacerbated by the fluctuations of wind speeds in the jet stream core. It works this way: The jet stream boundary between pressure systems can create two types of wind flow. In most cases, usually in calm weather and near-normal temperatures, an almost-straight-line jet stream stretches across a region. But if there are very large ridges and troughs along the jet stream—represented on a weather map by deep wavy lines with sharp peaks and dips, respectively— stormy weather ensues, dragging in frigid polar air in winter and heat waves in summer.

In the United States, the northern polar front jet stream is the strongest; it is only about 180 to 300 miles (290 to 480 kilometers) wide, and 1 to 2 miles (2 to 3 kilometers) deep. It normally has speeds from 100 to 150 miles (160 to 240 kilometers) per hour; in winter, however, the temperature difference the cooler and warmer air masses increases the speed of the jet stream, sometimes up to 300 miles (480 kilometers) per hour. Such speeds contribute to our saving airline fuel when flying in an eastward direction. In fact, flights across the Atlantic Ocean from the United States to Europe take full advantage of these strong winds, often cutting off one to two hours of flight time. Flight paths in the opposite direction avoid these strong head winds to conserve fuel.

Like anything else in life, there is also a downside. Much of the clear-weather turbulence felt in high-flying aircraft—the bumps that spill coffee on your lap and make trips to the bathroom interesting—are associated with fast-moving jet streams.

• *Jet stream information from the*
University of Illinois' Department of Atmospheric Science:
http://ww2010.atmos.uiuc.edu/(Gl)/guides/maps/upa/jetstrk.rxml

Jet stream clouds over Fraser Island, off Australia's east coast, September 1991. (courtesy of NASA)

• The Other Jet Streams •

Like water in a garden hose, the flow of our planet's jet streams is not static, and conditions are always ripe for some strange twists and turns. For example, the eastward-flowing polar and subtropical jet streams in either hemisphere can join, creating one powerful jet stream, an event that has occurred over China and Japan. Shorter, slower jet streams also pop up around the world, sometimes even flowing east to west, but they appear and disappear quickly. As with most weather patterns, the topography of a local area can affect a jet stream. Like a babbling brook flowing over stones, jet stream winds slow as they travel up mountains and accelerate as they run downslope. The faster-moving portion, called a *jet streak,* is responsible for producing many strong weather systems.

Even smaller jet "streams," called simply *jets*, can be found from 2 feet (0.6 meters) to 1,000 feet (305 meters) in altitude, but only at night. These low-level, localized jets—so far detected on the plains of the United States and in Canada, China, Australia, Argentina, and western Europe—form most often during calm conditions, as cool air sinks to the ground, displacing warmer air. The warm layer stays on top of the cool air, acting like a solid to create a path for any wind passing above. Air flows rapidly over this inversion, creating a jet with winds up to 60 miles (97 kilometers) per hour, which airplanes experience in the form of turbulence on clear nights. And if conditions are right, these sudden winds also help develop (or feed already developing) thunderstorms with moist, warm air.

2

Tricks of Light and Shadow

The term weird doesn't even fit when it comes to the tricks light and sky seem to play on us, most commonly the result of sunlight or moonlight reflecting, refracting, bouncing, or just disappearing. The legacy of this magical legerdemain goes back for centuries in which the displays inspired countless legends, from sunbeams and rainbows to flashes of green from the sun. One illusion attached to legend after legend is the mirage, often caused as warm air sits on cooler air, behaving like a huge natural mirror. This play of light often exposes places hidden by the curvature of the Earth. One incident occurred in the Orkney Islands off the coast of Scotland: snowcapped mountains, complete with a train and village below, appeared at the north end of an island. But there are no trains, mountains, or such villages on the islands. In this case, the vision was a mirage of Norway—a spot 500 miles east of the islands—that disappeared after sunset.

Sunbeams: Stairways to Heaven

Sunbeams have long been the favorite motif in nature and inspirational photography; *crepuscular rays*, light and dark beams fanning out from the Sun, are the most popular. *Crepuscular* is from the Latin *creper*, meaning "dusky" or "twilight," and the rays are also known by other names: sunbeams, Rays of Buddha, Jacob's Ladder, or Ropes of Maui. Years ago, our grandparents used to say that when the rayed beams reached the ground, the clouds soaked up water for the next rain.

Crepuscular rays form as sunlight illuminates *aerosols*, or liquid and solid particles, in the air. (The easiest way to understand the illumination of aerosols is to watch sunlight bouncing off dust particles after you've cleaned any room of your house.) The bright crepuscular rays appear when sunlight shines between broken or scattered clouds; conversely, the darker rays are shadows of the clouds cast on the particle-filled air. The counterpart of the crepuscular rays, the *anti-crepuscular rays*, can often be seen 180 degrees from the Sun. Just stand with your back to the Sun: The anti-crepuscular rays, if visible, will be a dim, fan-shaped rendition of the crepuscular rays on the opposite horizon.

Although all crepuscular rays appear to be radiating from a singular point behind a cloud, this is merely an optical illusion. The Sun's light reaching the Earth is essentially parallel, but it appears to converge at a far "horizon" point, similar to the way parallel railroad tracks seem to meet at the horizon line.

Crepuscular rays are caused by sunlight beaming off the aerosols in the atmosphere. They are most often seen as the Sun is rising or setting.

Another sunbeam seen at or just before sunrise or sunset, especially in the winter months, includes tall, bright vertical streaks of light above or below the sun called *light pillars, spikes,* or *Sun crosses.* They are most commonly a ghostly white, but depending on the color of the sunrise or sunset, can be vivid red, orange, or yellow. The pointed rays of light occur as sunlight reflects off ice called *diamond dust* high in the atmosphere, tiny sparkling hexagonal (five-sided) crystals that fall so slowly they seem to be floating in the air. Each flat surface of the crystal acts like a tiny mirror, reflecting the color of the sunrise or sunset.

Sun pillars are most often seen in the winter months, when the rising (or set-ting) Sun's rays reflect off the flat surfaces of hexagonal plates of ice crystals in the atmosphere.

Rainbows: Somewhere Over What Type of Rainbow?

Rainbows bring up thoughts of *The Wizard of Oz*, but there is really no way to "get over" a rainbow. The true arc of a rainbow can be seen only directly opposite the Sun. That is, you will see the rainbow only when the Sun is at your back and lower toward the horizon, so be wary if someone tells you they saw a rainbow when the Sun was very high in the sky.

When the sunlight hits an individual water droplet, it is *refracted*, or bent, as it enters the drop. This bending of the light separates the white light into its component colors, just like a miniature prism. This multicolored light then reflects off the back of the water drop, just like your image in a mirror, and heads back the way it came, toward the Sun. As the light leaves the water drop back into the air, it refracts again at a specific angle, and we see a certain color. Multiply this process by

Double rainbows are rare but have the same origin as a single rainbow: the refraction of sunlight on the water molecules after a rainstorm. Rainbows are most often seen in the summer. Here, the rain falling from the edge of a storm is seen as a light veil of the primary bow; heavier rain is falling on either side of the secondary bow.

the millions of raindrops falling on a distant horizon, and a rainbow bursts into color—violet, indigo, blue, green, yellow, orange, and red from the inside to the outside of the arc. Because the angles to and from the raindrops are so critical, no two people ever see the exact same rainbow. And sometimes, when conditions are favorable, some of the sunlight bounces twice inside the drops, producing a larger, fainter secondary rainbow above the main one. The color sequence in this secondary rainbow is reversed.

There are also rainbow rules: Since a rainbow is created by the interaction of the Sun's rays and water droplets, the resulting bow would actually appear as a large circle, except for the interference of the Earth's surface that only allows us to witness, at most, a half circle. When the Sun is more than 42 degrees above the horizon, the primary bow cannot be seen because of angle effects; if it is more than 51 degrees above the horizon, a secondary bow cannot appear.

One other rule, but mostly folklore: Rainbows may tell you more about the surrounding rainstorms than you thought you knew. If the rainbow is mostly green in color, the rain will continue to fall. If red is the dominant color, winds will accompany the rain. And finally, if blue is the brightest color, you can count on the air clearing.

• Cloudbows and Fogbows •

As in a rainbow, when sunlight bounces off the tiny water droplets in fog, you can see a white *cloudbow* (or white rainbow). More colorful *fogbows* form through a similar process, but in this case, they are usually associated with fogs containing larger water droplets. The best time to catch a cloudbow or fogbow is early morning, when the sunlight begins to burn off and break up a heavy fog.

• For more about cloudbows, try linking to:

http://www.cloudbow.com

Halos: Lord of the Rings

Some of the showier weather phenomena—in particular, the rings seen around the Sun and the Moon—are short-lived. The angled light from these two bodies plays with the particles in the atmosphere, mostly water droplets and ice crystals, to create these phenomena.

For instance, the complete, glowing circular disks around the Sun or Moon are called *coronas* (crowns). These may be one or a series of colorful disks enveloping and extending beyond the Sun or Moon, usually around a radius of 10 degrees, or about a finger's width away from the outside of the Sun or Moon. They are seen mostly in association with the Moon, as the bright Sun often washes out its ring. The light rays passing through relatively thin clouds containing small water droplets of a fairly uniform size are *diffracted* (scattered) into diffuse bright, colorful circles. The process is similar to the scattering of blue light by the atmosphere's water and other molecules, forming the sky's color, but on a more localized scale. Under the best conditions, a corona can be even more bright and colorful than a rainbow.

While a corona seems to puff up the Sun or Moon, in contrast, the more common *halos* show up as distinct rings around these heavenly bodies. More often associated with the Sun, halos need a sky filled with high-altitude ice crystals like those accompanying an advancing warm front. Reflected light forms a perfectly white halo. But if light rays enter the crystals' top faces, they refract into a spectrum of color, then exit by one of the six side faces. The sum of all the light refracted by the numerous ice crystals produces a colorful ring, or halo, at an angle of 22 degrees—two finger widths—from the Sun or Moon. Halos' color sequence is the reverse of coronas, because of refraction in ice crystals rather than water droplets. And if conditions are just right, smaller and larger halos appear—as small as 9 degrees and as large as 46 degrees. One other type of halo only recently understood is the *elliptical halo,* an oval halo around the Sun that appears to originate from clouds in the middle levels of the atmosphere.

Not all halos reach completion, and these fragments may resemble bright lights on either side and above the Sun, creating a strange illusion of three Suns in the sky. These *parhelia,* also known as *Sun dogs* or *mock Suns*—and around the Moon, *paraselenae* or *Moon dogs*—are produced under the same conditions as halos. The sunlight passes through a high, thin layer of ice crystals. But in this case, the tiny, thin, six-sided plates of ice need to have their flat crystal faces oriented along a horizontal plane. And all this has to happen while the Sun is 45 degrees or less above the horizon. Some of the most spectacular Sun dogs and halos occur in Antarctica, where atmospheric ice crystals are more common and the horizon is wide open. In the milder climates, the appearance of a halo or Sun dog often signals the presence of high cirrus clouds made of ice crystals, which to farmers (and those of us who watch the weather) often means rain is on the way.

• *For more information on optical phenomena, try linking to the Finnish page:*
http://www.funet.fi/pub/astro/html/eng/obs/meteoptic/links.html
and for more general coverage:
http://www.atmos.uiuc.edu/wxinfo.html

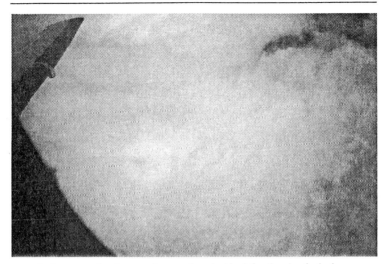

A glory can often be seen from a plane: Just look for the plane's shadow, with a circular rainbow around it, on a cloud below.

• The Wilder Side of Halos •

The next time you're flying in an airplane and are bored by the in-flight movie, look out the window down toward a cloudbank. If you are traveling in the right direction, so the sun is behind you and the plane, you might see an optical phenomenon called a *glory, anticorona,* or *Brocken bow.* This is usually a small series of rainbow-colored circles surrounding the actual shadow of the plane and is caused by the reflection of water droplets in the cloud.

Strange rings seen around the Moon, called *lunar mock halos,* appear as a collection of rings—some full, some semicircles, and all "hooked" to one another like a ring puzzle—surrounding the bright disk of the Moon. The multiple rings are rare and are most often seen in the colder regions of the world, as the air becomes filled with an ice-crystal haze.

Sunset Shadows: The Shadow Knows

Sunset in most places of the world holds more than the cooling of temperatures. Sunset is also a time of *shadows*, and in particular, shadows of the Earth. As the colors of twilight fade in a clear, cloudless sky, Earth's shadow rises in the opposite side of the sky, or in the east. To an observer on Earth, the shadow looks like a huge dark band along the eastern horizon, best observed if the horizon is unobstructed, or if you are on higher ground with no obstruction. This *twilight wedge* broadens as the Sun sets, the upper part turning into a dusky gray. From a high-altitude jet's-eye view, the wedge actually looks like the sunlight spilling over Earth's curved horizon, its bottom cut off by the planet's shadow. And as you and the Earth's surface spin more toward the darkness of night, the band gradually disappears.

Of course, space shuttles have an even better view of the huge Earth shadows. One of the more fascinating shadow effects comes not from clear skies but from tall, massive thunderstorms. As storms grow in size (especially over the Tropics), and the Sun is close to setting or rising at that point on Earth, the shadows of the thunderstorms create dark stripes across the planet.

Green Flash: Not Your Average Superhero

Don't think the *green flash* is something from a cartoon action adventure. The flash is a phenomenon caused by a play of

---

• Please Heed This Warning! •

WARNING! DO NOT OBSERVE THE SUN WITH THE
NAKED EYE, BINOCULARS, OR TELESCOPE WITH-
OUT THE PROPER VIEWING EQUIPMENT. OBSERV-
ING DIRECT SUNLIGHT CAN CAUSE LASTING
DAMAGE TO YOUR EYESIGHT.

---

optics in the atmosphere, a flash of green on the horizon right at the point where the Sun sets.

Don't expect to see it everywhere. But if there is a locale with a clear, cloudless, open horizon, you may be able to catch a glimpse of it. As the Sun sets, the atmosphere refracts (bends) sunlight, creating a spectrum of color. For a brief moment, red, orange, yellow, and finally green appear in rapid sequence; as the Sun rises, the colors appear in reverse. In reality, the green flash should be renamed "blue flash" or "violet flash," since the refraction is actually greatest toward that end of the spectrum. Particles in the air and water vapor are the culprits here. The particles eliminate the blue and violet light, and the water vapor eliminates the yellow light, leaving primarily green and red.

Green flashes are best seen at sea, on a hilltop, along western-facing coastlines such as the California coast, or generally any sparsely populated horizon line where you can observe the very last sliver of the Sun as it slides into the sunset. In the higher latitudes during the summer, the Sun seems to be setting slower, and a green flash may last many seconds. Along the poles, as the Sun skims the horizon at various times of the year, the flash may last many minutes. In fact, during Admiral

Byrd's expedition to Little America in 1929, a green flash was seen on and off for thirty-five minutes. But for the majority of observers in the middle latitudes, the flash more commonly lasts only a few seconds.

Whether or not the phenomenon is common seems to be a matter of debate. Some astronomers say that green flashes are rare, sighting them only once out of a hundred sunsets; others say they are actually common, especially over still oceans or lakes with a clear and unobstructed horizon. Indeed, if the Scottish saying is correct—that anyone who has seen the green flash will never err in matters of love—then everyone should have a happy love life, if the phenomenon is so common. No, the problem is not with rarity. It truly happens each time the Sun sets. The trouble is in trying to observe it.

You can behold an additional sight after the Sun sets: The *Novaya Zemlya effect*, named after the chain of islands extending from Russia's Ural Mountains into the Arctic Sea. Just above the horizon, the light from the Sun's face is bent into thin slits, like a stack of plates forming a pyramid. You can best view the effect at higher latitudes, where the angle of the setting Sun and horizon is very shallow. But you can also occasionally observe it several minutes after sunset in the middle latitudes.

UFOs and Nighttime: Unsteady Space

It doesn't matter if it's summer or winter. The public reports UFOs, or unidentified flying objects, to planetariums and observatories any time of the year, and for good reason. In the summer months, more people are outside on vacation or just taking walks in the cooler nighttime air, thus more people are watching the nighttime skies. In the winter months, clear skies make the stars seem brighter and more eerie. All these conditions invite reports of strange objects in the sky—strange at least until the objects are identified.

The characteristics of the local atmosphere are to blame for the frantic calls. The differences in temperature and humidity in an air mass tend to break starlight into dancing specks of light that our eyes interpret as twinkling. This is especially true along the horizon, where certain bright stars sit, blinking madly through the turbulent lower layers of the atmosphere.

The planets aren't exempt, either: Venus and Mercury, known as the morning or evening stars, appear as points of light on the horizon, Venus occasionally shining brightly enough at night to cast a shadow of objects on Earth's surface. Especially in their crescent phase, both planets shine almost as bright as the most luminous stars. And because the planets' orbital paths never take them too high in Earth's evening or morning skies, the thicker atmosphere and heat from buildings along the horizon cause the planets to waver and blur.

During all these times, people claiming to have seen UFOs deluge police offices and observatories with phone calls. But if they watched closely, the callers would notice one major clue: The UFO twinkles but doesn't budge an inch.

Mirages: Seeing Things

Stories of the desert always seem to include the ubiquitous *mirage* in the distance, beckoning the hero to water or food, but he finds only sand at the end of his journey. In reality, a mirage as an illusion is a misconception. There really *is* a physical object in the distance when you see a mirage. Tricks of temperature and light in the atmosphere change and refract the object's appearance and location.

A mirage, which can be viewed almost anywhere, from deserts to the oceans, occurs when light is refracted (bent) as it travels through the atmosphere. As the light bends, the image of the actual object is displaced. As with all mirages, temperature is the reason for the double or multiple images of the object. If the air temperatures near the ground increase, the image is displaced downward, forming an *inferior mirage*; if the air temperatures increase with height, the image is displaced upward, forming a *superior mirage*. The curved light paths cause a distant image—from a town across the desert to a boat on the ocean's horizon—to appear slightly displaced from where the actual object lies, a phenomenon you can also see through a telescope or binoculars.

You have probably seen a mirage. The most common ones appear as waterlike spots on a distant horizon over heated asphalt roads on a hot summer day, with the water miraculously "evaporating" as you approach in a car and reappearing in the rearview mirror. The highway mirage is caused by the heating of the road surface, which raises the temperature by tens of degrees above the surrounding surface air. The warmer air thus refracts the light upward. What looks like a pool of water is actually light from the sky refracted into your line of sight (inferior mirage). The shimmering you see in association with a mirage is due to the constant movement of the atmosphere, the same turbulence that causes stars to twinkle at night.

For the more intense temperature differences, extreme mirages can form, including the *hafgerdingar effect,* more commonly called *fata morgana* ("Morgan fairy," named after the evil witchlike creature who built fake castles in the air to fool legendary heroes). Here, temperature layers are not uniform, breaking the superior mirage into fragments. The mirage is mainly in the shape of spikes and shoots, hence its other name, "castle in the sky." Such temperature extremes also form the *fata bromosa* ("fairy fog"), which appears as a horizontal white band over the ocean waters in the Arctic regions, caused by superior mirages over the snowfields. Mirages can even appear in the winter in the more temperate regions, especially when cold air lies over sun-warmed asphalt. And at night, headlights from a faraway car can cause a mirage just before the ground cools down.

• Long-Distance Mirage •

Mirages pop up over very long distances, especially if the light is trapped in a cool layer of air between two warmer layers. Such a mirage was seen in 1939: The mirage of Snaefellsjökull in Iceland was seen from the ocean at a distance of 342 miles (550 kilometers) away, on record as the most distant object seen in a mirage, according to the *Guinness Book of Records.* In fact, it's said the people of Greenland knew about North America long before its "discovery" by the European explorers. When the temperatures were just right over the ocean, it was easy to see a mirage of the huge continent.

• *For more information on optical mirages and the Mirage Project in the United Kingdom, try linking to:*
http://www.lapr.demon.co.uk/mirage.htm

3

Weird Cloud Formations

Our attempts to see the land below the highest mountain peak along the East Coast, Mount Mitchell in North Carolina, have often been foiled in the past by thick fogs. It makes sense: Fog on the mountain occurs an average seven out of ten days, as the huge mountain seems to grab any sort of cloud that comes its way. But with luck, and beating the statistics, we were finally treated to the view on a clear, windy day. From the tall observation post, we watched as nearby clouds became caught in the eddies and swirling winds trying to get by the mountain and its range. One cumulus cloud treated us to several curling tendrils, often seen in the mountains as the winds carve the clouds, while another small cloud seemed to form a stack of plates high in the air. For those people who noticed, the cloud looked bizarre, even alien. No wonder so many people report these mountain clouds as UFOs.

Clouds: Wave Hands Like Clouds

All of us have no doubt taken time to lie in the summer grass to gaze at fair-weather clouds. We see bunny rabbits, frogs, or even faces in the clouds, as well as clouds that defy any familiar shape. Formed by the winds and tempered by moisture and temperature, clouds are familiar to all of us. These continuously rejuvenated collections of very tiny water droplets or ice crystals suspended in air are a vital part of our life cycle and bring much-needed water around the world.

Clouds form as warm, moist air rises, cools to its dew point, and condenses. Because of their minuteness, the water droplets can be suspended in the atmosphere by the upward movement of air. And even if no updrafts exist, the small size of the droplets makes them drift downward very slowly, so slowly that they quickly evaporate and return to the vapor state. Then they rise again, condense, and drift downward in a cycle. The localized area where these collective droplets are visible is called a *cloud*.

The majority of clouds occur within the *troposphere*, a layer of the atmosphere that reaches about 7 miles (11 kilometers) in height, just a little higher than long-distance planes fly. Clouds are normally white but can also range from almost black to any shade of gray. The tiny droplets of liquid scatter all the colors of the sunlight, resulting in white clouds. When they are thin enough—that is, the droplets are not as numerous, or are farther apart—the whole cloud looks white from the transmitted light, like the puffy clouds you see on a summer's day. As the number of droplets increases and the cloud becomes thicker, less light can pass through to the bottom; from our vantage point on the ground, the cloud has a gray appearance typical of a rainy day. If the cloud is thick enough, it might even appear almost black, typical of the underside of thunderstorm clouds. The coloring, of course, is all a matter of perspective: From an airplane, these "dark" clouds would be bright and white, as the clouds reflect the sunlight. Similarly, when the rays of the sun strike the clouds at just the right angle, usually at sunrise or sunset, other colors such as red, pink, and orange are quite visible.

The overall classification of clouds is based on a system devised by Luke Howard, a nineteenth-century London

Cumulus clouds getting caught on the windward side of the Presidential Range, New Hampshire. Note how the clouds dropped snow as they rode up the mountains.

Seen from a vantage point in the Blue Ridge Mountains, a line of cirrus clouds fills the western sky, signaling the advance of a warm front.

apothecary. His three main categories derived from shape: *cirrus*, from the Latin for "lock of hair," the wispy high-level clouds; *cumulus*, from the Latin for "heap," for the lumpy masses of clouds closer to the surface; and *stratus*, from the Latin for "layers," for the clouds that lie horizontally. Other Latin cloud names are used in combination with the three main categories. For example, lumpy, showery thunderstorm clouds are cumulonimbus (nimbus meaning "shower"), and similarly, showery layered clouds are nimbostratus.

- *Try the Center for Clouds, Chemistry, and Climates at the University of California, San Diego:*
  http://www-c4.ucsd.edu/

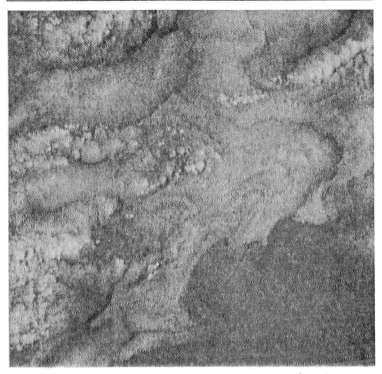

A cloud cover over the Atlantic Ocean is a strange sight as seen from the space shuttle mission STS-59.

• Your Very Own Cloud •

To get a close-up and personal look at a cloud, albeit on a small scale, pop open a bottle of soda. The liquid inside the bottle contains dissolved carbon dioxide. This gas creates pressures more than twice the amount you feel around you when you stand at sea level. As you uncap the bottle, the gas rapidly escapes and the pressure drops precipitously. The space above the soda contains mostly carbon dioxide, but it also contains water vapor (after all, the soda is mostly water). So as you pop the cap, the gases within the neck expand rapidly; as a consequence, temperature in the neck also drops precipitously. The water molecules become sluggish, allowing some of them to form small clouds as water condenses—your own personal cloud show for a few seconds.

Mountain Clouds: Sky Sculptures

Wind flow over the peaks of mountain ranges creates all sorts of strange cloud formations. Mountain wind patterns known as rotor and lee waves sculpt the clouds into wild shapes. The rotors, or eddies, form as winds blow off the mountain tops, creating roll-shaped cumulus clouds.

But it's the lee waves that mold the more fascinating clouds. *Lenticular clouds* resemble smooth, flattened almonds, forming just above or to the lee of the peaks. These large lens-shaped clouds have often been reported as UFOs by tourists visiting in the mountains. They are dramatic sights over the huge mountain ranges worldwide—from the higher Himalayas, Rocky Mountains, and Andes—and even the "shorter" peaks of the Appalachians, Ethiopian Highlands of Africa, and the Sikhote Alin Range of Russia.

In reality, lenticular clouds form when high-speed winds in the middle layers of the atmosphere flow over the mountains, creating air waves on the lee side, the so-called mountain wave effect. Where these winds rise, clouds form, following the sinuous path of the waves; where the winds fall, the clouds dissipate. If the mountain range is fairly regular in shape, with the winds blowing steadily at right angles to the mountains, a more regular pattern will develop. If the air above the moun-

Lenticular clouds form just beyond the lee side of a mountain; This forming lenticular cloud is typical. Such clouds have often been reported as UFOs when fully developed.

tains includes alternating layers of dry and moist air, the resulting lens-shaped clouds resemble huge stacks of dinner plates—and, to the uninitiated, a UFO.

Cap clouds, or foehn walls, appear as seemingly stationary clouds hovering over the tops of mountain peaks. Found most often along the north–south line of the Front Range of the Rocky Mountains in the United States, they are long, low clouds just behind the peaks (to the west, in this case). Traveling by train toward the Rocky Mountains is the best way to view these clouds: Even from great distances, the mountains often appear taller than they really are, the foehn wall of clouds adding to the peaks, giving them a more gigantic appearance.

An even more bizarre form of cloud includes the *Kelvin-Helmholtz*, a row of small, curving clouds resembling the ocean waves we all used to draw as young children. The crest-

This storm in the Blue Ridge Mountains of North Carolina shows the effects of the mountains on the clouds, even during a storm. The twisted cloud is similar to the Kelvin-Helmholtz curving clouds found around the lee side of mountains.

ing waves form as strong winds increase across the top of the clouds, sculpting them into wave crests running to the lee side of the peaks. If you're lucky and catch these clouds at sunrise or sunset, they glow with brilliant colors, the setting sunlight bouncing off the wavy clouds. .

High Clouds: A Technicolor Show

Oceangoing vessels have an advantage: wide horizons, with little to come between that horizon and the eye. Ships traveling past the Falkland Islands in the Southern Hemisphere's summer or heading through the pack ice in midwinter often witness two of the rarest clouds in the world: *noctilucent* and *nacreous clouds*.

Rare noctilucent clouds are blue to silvery-colored clouds (though they can sometimes have a reddish-orange hue), often mistakenly identified as lenticular clouds. Seen mostly in the warmer summer months at twilight, they are also called night-shining clouds. Brightly lit long after the Sun sets, the clouds are believed to form from the reflection of ice surrounding meteoric dust. Another theory states the clouds may reflect increasing concentrations of water vapor in the upper atmosphere.

Amazingly, because of their heights of 47 to 56 miles (76 to 90 kilometers)—and contrary to most clouds in the lower-level troposphere—noctilucent clouds are viewed several hundred miles away from the source. For instance, clouds forming over Scandinavia would be visible from Scotland too. They are viewed only between 50 and 75 degrees north latitude in the Northern Hemisphere and between 40 and 60 degrees south latitude in the Southern Hemisphere.

The lower nacreous clouds—about 12 to 19 miles (20 to 30 kilometers) in height—are called mother-of-pearl clouds because of their seashell-like iridescence. The silvery clouds, which resemble thin, feathery cirrus clouds, can form and change quite rapidly. And as with many rare phenomena, there is no satisfactory explanation for the clouds. Seen during the winter months, when the Sun is low, these clouds remain brilliant long after sunset or before sunrise. The clouds seem to be more prolific when there is a deep west-to-northwest airflow. Some people have observed the clouds around large mountain ranges, fueling speculation that, because they show little or no movement, the clouds may be similar to static mountain clouds but on a much grander scale.

Fog: Now You Don't See It

Fog can be thought of as a cloud on the ground. In other words, there is no discernible difference between flying through a high cloud in an airplane or driving through fog on your way to work. The mechanism of formation is the same: Warm, humid air is cooled to its dew point by colder air, and condensation occurs.

Take, for instance, the warm land breezes passing over the cool waters of the Labrador Current, just off the New England coast of Nantucket, which lead to some of the foggiest coastal conditions anywhere in the world. This area, notorious for fog, may also have contributed to the loss of forty-eight passengers and the sinking of the 29,000-ton Italian liner T/N *Andrea Doria* on July 25, 1956; she was a ship traveling from Genoa, Italy, bound for New York City. Late that night, some 45 feet (14 meters) of the bow of the MV *Stockholm*—a Swedish-American ship that had departed from New York City that morning—penetrated the starboard side of the *Andrea Doria*. After lying on her side for about eleven hours, the *Andrea Doria* sank 250 feet (76 meters) beneath the surface of the ocean. The true reason for the collision is still a mystery, since both ships were equipped with the most advanced radar equipment of the time, which enabled them to search through zero-visibility fog.

Back on land, fog most often forms on clear nights with no wind. On top of hills, the heat radiates away; the cooler ground then chills the air above it. This cold, denser air flows down into the valleys, where it cools warm, humid air to its dew point, thus condensing the water vapor into tiny droplets to form the fog. (If the layer of condensation is thinner, a dew will form; if it is thicker, a fog will form.) This is also called *radiation fog*, so named because it is associated with radiation cooling of the ground. As with most fogs, it will disperse as the Sun's rays gradually warm the ground; the unevenness of heating on the ground is why the fog will disappear in one place but take longer to dissipate in another. If this type of fog consists of tiny particles of ice instead of water droplets, it is called an *ice fog*.

Advection fog also forms from condensation, but it is not caused by a change in ground temperature. In this case, warm, moist air drifting into cold air (or cold air into a warm, moist environment) causes the fog on land or over the ocean. *Sea fogs* off coastlines are some of the best examples. For example, fogs form over the North Sea along the British Isles, as air from warmer ocean currents drifts over colder ocean currents. *Valley fog* is also an advection fog. The cooler, denser air drains into the valleys from the hilltops, causing a billowy, cottony fog to cover the lowlands.

Another often misunderstood obscuring phenomenon is *mist*, the term given to a very thin fog. The only difference between a mist and a fog is that you can see farther in a mist. This term is also popularly, but incorrectly, used to describe a fine drizzling rain.

Far out to sea, sailors have seen something easily mistaken for an unknown phenomenon, but in this case, it's not a mirage. It is called *sea smoke* or *steam fog*, a number of towering tufts of what looks like smoke billowing above the ocean horizon. The smoke is truly a fog, formed much in the same

Ground (or radiation) fog is actually a stratus cloud filling the valley. The wispy and uneven fog shows that the Sun's rays are warming the ground and dispersing the fog.

way as fog on land—from a significant difference between the air temperature and the sea surface temperature, usually at least 10 degrees, with the sea surface being warmer. Areas where currents meet, such as the warm Gulf Stream meeting the cooler Arctic waters, are prime areas for such "smoke."

• The Dark Side of Fog •

Not all fogs are innocuous masses of water particles. Some are downright deadly. Such harrowing fogs occurred in the city of London during the 1950s, when sulfur dioxide from burning coal combined with the natural fog to produce a deadly mix of soot and sulfuric acid droplets. In one case, in 1952, a stagnant air mass during a four-day period caused a massive buildup of particulates and sulfur dioxide so thick it was often difficult to see objects inches away. Everything virtually came to a standstill, and when the air cleared, 4,000 were dead, with more than 100,000 ill from the effects of the black-and-yellow fog. The British Clean Air Bill of 1956 all but eliminated the possibility of another such fog, with soft coal use largely banned.

The development of such fogs is not the only potential danger. There is also smog, the term apparently coined in 1905 by British physician Harold Des Voeux to describe a combination of smoke and fog. In reality, smog is neither smoke nor fog but a chemical brew found mostly near major cities and industrial centers. Smog is actually the result of a photochemical reaction: Sunlight reacts with the exhausts from automobiles, individual household machines—even chain saw motors produce smog-forming chemicals—and power plants, factories, and industrial emissions, especially oxides of nitrogen and volatile organic compounds. The resulting smog carries a plethora of components, including ozone (different from the upper atmosphere's ozone formed naturally by a photochemical reaction between sunlight and oxygen, not pollutants) and the minute solid particles that make the smog so visible.

Environmental controls have cleaned dirty fogs and smogs in many industrialized nations, but smog remains difficult to control, especially in larger cities. Other potential, and beginning, producers of smog include the many developing countries, places that don't have the funds to maintain high air-quality standards.

4

Rainstorms, Snowstorms, and Thunderstorms

We all know the "weekend effect": It's summer, and it's been raining all weekend. At 6 P.M., the clouds seem to split apart, revealing the Sun sinking lower on the horizon. The day is almost over, and thanks to the rain, you haven't finished half the chores—or played half the time—you wanted. Monday rolls around and the sunshine almost blinds you as you travel to work. In fact, we once counted thirteen rainy weekends in a row. Is there something to this weird periodicity? Scientists have found a possible reason: us. According to recent studies, industrial activity—our weekday work—may create this weekend scenario. One suggestion is that the particles sent into the atmosphere from cars, trains, factories, and sundry human activities allow water vapor to condense into raindrops. This artificial "seeding" is not an excuse to get rid of our weekend breaks, but it's an example of how we can create our own rainstorms.

Rainy Days: Rain, Rain, Don't Go Away

We live in an atmosphere with normally less than 100 percent water vapor. This is a good thing too, or evolutionarily, we would still be in the fish phase. And contrary to popular belief, if the relative humidity is 100 percent on the ground, it will *not* be raining.

There is not a spot in the world that does not experience some type of precipitation. About 90 percent of all the water vapor in the atmosphere comes from evaporation of ocean water, with the rest from lakes, streams, and rivers, all the result of the Sun's rays heating and evaporating the waters' surfaces. Warm, moist air is lighter than dry air, and thus rises. If the atmosphere is unstable—when the natural restlessness of the atmospheric winds causes the air to move horizontally, vertically, or both—the humid air continues to rise to great heights. If the air rises high enough, it cools, and temperatures fall to the dew point; the water vapor reaches a saturation point, condensing and forming clouds. Finishing off the cycle, the water vapor precipitates out in the form of tiny droplets of water (or ice crystals, depending on the surrounding air temperature). The small condensing droplets gather together to form larger particles, until they are big enough to overcome the upward movement of the air in the cloud, eventually to fall as rain.

From there, a multitude of things can happen to the tiny particles on the way to the ground, depending on local conditions. The droplets may evaporate before reaching the surface, especially in dry climates. They can merge to form larger raindrops, striking the surface as rain, drizzle, snow, ice, and other forms of precipitation. Or they can remain suspended in the sky. In general, rain comes from certain types of clouds. Nimbostratus clouds, which form at fronts, result in *frontal rains;* cumulonimbus clouds, which form because of convection, usually result in *convectional rains,* providing intense, though short-lived, downpours.

What is the official difference between a typical rainstorm and a severe storm? Oddly enough, lightning is not the key. Meteorologists define a severe storm as one that creates either hail larger than three-quarters of an inch (1.9 centimeters) in diameter, winds greater than 57.4 miles (92.4 kilometers) per hour, or both.

As for the actual raindrops themselves, images we see in cartoons don't even come close to real life. In other words, rain does not fall as perfect spheres or in teardrop shapes. Drops smaller than 0.08 inch (0.2 centimeter) in diameter are spherical as they fall. Larger drops have flattened bottoms and bulging sides, like a half of a burger bun. And drops larger than a quarter of an inch break up into smaller drops as they fall.

Speeds vary among raindrops too. The smallest raindrops usually never reach the ground; instead, they continue to be pulled up and down by the air shafts within a cloud. The largest raindrops, on the order of about a quarter of an inch (0.65 centimeter) in diameter, fall to the ground much more rapidly, up to 16 to 20 miles (26 to 32 kilometers) per hour. Snow, the lighter, frozen version of rain, falls at a gentle 1 to 5 miles (1.6 to 8 kilometers) per hour. Another way of looking at rainfall rates is by overall accumulation. Light rain falls at the rate of 0.1 inch (0.25 centimeter) or less per hour, with individual drops seen. Moderate rain falls at 0.11 to 0.30 inch (0.28 to 0.76 centimeter) per hour, and heavy rain falls at more than 0.30 inch (0.76 centimeter) per hour.

• Rainfall Records •

Rainfall records break every year. Here are some of the latest:

• It's hard to believe that somewhere they call paradise is one of the rainiest spots in the world. That place is Hawaii's Mount Waialeale, a mountain with an average of 360 rainy days per year.

• But Mount Waialeale is still not the wettest place on Earth. The record goes to Mawsynram, Meghalaya, India, a place that averages 467.5 inches (1,187.45 centimeters) of rain per year, with Tununendo, Colombia, South America, running a close second, measuring an average of 463.4 inches (1,177 centimeters) of annual rainfall.

• There are, to be sure, many places with the dubious distinction of being very dry. Currently the driest measured location on Earth is the Atacama Desert, in Chile, South America. There is virtually no rainfall, only a passing shower several times per century, yielding an average of 0.003 inch (0.08 millimeter) of rain annually.

• Another rainfall record is in Unionville, Maryland: a total of 1.23 inches (3.12 centimeters) of rain fell on July 4, 1956. It may seem like nothing out of the ordinary, since the state sits in a notably rainy, humid region of the mid-Atlantic. But in this case, the rain came in only sixty seconds, so far the United States' one-minute record rainfall.

• There are global rainmaker records too. In 1942, Cherrapunji, India, in the Himalayas, experienced 1,042 inches (2,647 centimeters) of rain, with 364 inches (925 centimeters) falling in one month.

- In 1952, between March 13 and March 18, the world's five-day record rainfall was set in Cilaos, Réunion Island, in the Indian Ocean, as a tropical cyclone produced 151.73 inches (385 centimeters) of rain. But another world record was broken at the same time: A total of 73.62 inches (187 centimeters) of rain fell within a twenty-four-hour period in Cilaos.

- And finally, one of the fastest rainfall records: In sixty minutes on August 1, 1977, Muduuocaidang, Inner Mongolia, experienced 15.78 inches (40 centimeters) of rain.

Flash Flooding: More Than Enough Water

The July 31, 1976, celebration in Colorado marked the state's centennial. It was an unusually windless night, allowing a

A line of liquid precipitation, or rain, falls from a nimbostratus cloud, carried along with a huge thunderstorm system.

thunderstorm to develop and virtually sit over the festivities. The storm dropped from 13 to 15 inches (33 to 38 centimeters) of rain in a five-hour period, resulting in a massive *flash flood*. Much of the runoff fell into a narrow canyon that carried the Big Thompson Canyon River, producing a 20-foot (6-meter) wall of water that swept through the canyon in less than half an hour. Thousands of people—mostly campers visiting the area at the foot of Rocky Mountain State Park—were trapped by the rising water. People climbed the sides of the canyon as water carried boulders, trees, cars, mobile homes, and mud along the 25-mile (40-kilometer) stretch of canyon. At least 139 people and an entire town were swept away, and hundreds more were injured. It was a miracle that more were not killed. After all, flash-flood warnings were issued well in advance of the rains, but for the most part, they went unheeded.

Flash floods are just what the words imply: The flash speaks to the quickness of the event, the flooding to the increase in rainwater output. During such a flood, the water level of streams, or even water over land, dangerously rises, usually more than an inch (2.54 centimeters) an hour, as if an unseen hand were wringing out the clouds. And they can happen virtually anywhere. For example, the southwestern desert of the United States is an area of infrequent rains, but during a cloudburst, the dry river channels swell with rapidly flowing rainwater. Other evidence includes cloudburst tracks seen in places such as Alabama, as flash floods accompanied tornadoes in 1872, washing out gullies 60 feet (18.3 meters) wide and 3 to 4 feet (0.9 to 1.2 meters) deep. The Big Thompson Canyon River flood represented a classic "wall of water," in which the runoff water funnels rapidly down gorges. Other rapid rises include relatively flat areas such as the plains, or urban areas where rainwater rapidly runs off extensively paved areas, collecting in street intersections, underpasses, and dips in roads.

Flash floods aren't always caused by just the rains. During times of excessive spring rains, rising river waters can create ice jams. Even a man-made or natural dam failure can precipitate a menacing rapid flood, most often initiated by excessive rains near the dam area.

Cold Weather Storms: Sorting Out Winter Weather

High above the warmer equatorial climates in the more temperate zones, rainstorms in the colder months of the year squeeze out plenty of precipitation, most of it frozen. Some of these storms are merely a nuisance, while others have the potential to cripple entire regions of the country.

The most damaging and treacherous storms drop strange forms of ice. *Graupel*, or soft hail, consists mostly of a mass of frozen cloud droplets and is common in snowy blizzards. *Sleet*, or ice pellets, are frozen raindrops that fall and make roads and walkways almost unpassable. Supercooled droplets, or those that remain liquid even though the temperature drops below freezing, strike an already frozen ground, forming a *glaze* of ice. *Hoar frost*, or "white frost," forms when the water vapor in the air changes into ice crystals, or dew freezes as it covers the colder ground. Its much harder cousin is *rime ice*, which forms a white, milky deposit of ice as supercooled water hits a below-freezing object—the ice you hear about when a plane needs de-icing.

The typical, and often devastating, *ice storm* ("silver thaw") develops in the winter months in many middle-latitude regions as warm, moist air slides up over a colder, denser air mass. The resulting precipitation falls as raindrops mixed with ice crystals. The harbinger of the ice storm is usually sleet mixed with liquid drops. When this liquid hits the colder air at ground level, it congeals on objects below the freezing point. This often causes chaos as the weight of the heavy ice snaps trees, wires, vegetation, and other structures, and roads and sidewalks become caked with ice. In most cases, the warm air replaces the cold air, melting the ice. But in some places, such as southeastern Canada and the northeastern United States during the winter of 1997 to 1998, the warmth did not come. The ice remained, causing power outages everywhere, hanging on for weeks as cold arctic air stayed in place.

But overall, the best-known winter weather remains *snow*, precipitation that forms directly in clouds. Water vapor is deposited on a speck of ice or some other minute particle in the upper parts of the clouds, forming, on the molecular level, a six-sided crystal. What we see are needles, flat plates, tiny fingerlike crystals, or skinny columns, depending on the temperature within the cloud. The snowflakes we catch in our mittens are made up of many hundreds of these icy shapes. The snowflakes fall en masse from the snow-filled clouds similar to thunderstorm clouds, creating localized snowstorms.

A *blizzard* is another type of snowstorm, one that can blanket a region and, directly or indirectly, can paralyze an entire country by halting most forms of transportation. In such a storm, sustained winds blow greater than 35 miles (56 kilometers) per hour. Snow falls in blinding proportions, with visibility reduced to less than 1/4 mile (0.4 kilometer). In most cases, blizzards form after a winter frontal storm passes by, dropping less than 6 inches (15 centimeters) of snow. Then the winds switch from the south or southwest to the west or northwest, creating the blizzard conditions. The wind gusts produce massive drifts, sometimes over 10 feet (3 meters) high, usually on the lee side—opposite the windward side—of buildings or structures.

The East Coast of the United States has had its fair share of blizzards in the past years. The March 1993 blizzard shook the coast, causing $3 billion to $6 billion in damage and approximately 270 deaths. In February 1994 a southeast snowstorm struck, dropping ice and snow, causing $3 billion in damage and 9 deaths. But nature never seems to be finished. In January 1996 came "the Blizzard of 1996," so named as very heavy snow fell over the Appalachian Mountains, mid-Atlantic, and northeastern United States. The area was frozen by mountains of snow, followed by something just as bad—severe flooding due to snowmelt and more rains, which caused an estimated $3 billion in damage and 187 deaths.

One event often associated with a blizzard, but sometimes seen in a regular, gusty snowstorm, or even during an ice-crystal fog, is a *whiteout*. This phenomenon is marked by a high density of airborne snow, restricting visibility to a few feet or less, often making it difficult for drivers, skiers, or any other travelers to get their bearings.

• Snowflake Copies •

Contrary to popular belief, lookalike snowflakes *have* been found. The first identical snowflakes were found on November 1, 1986, on an oil-coated glass slide that was carried on an airplane flying over Wisconsin. The plane was flying at 20,000 feet (6,134 meters), and the temperature was −9°F (−23°C). The side-by-side crystals were shaped like rectangles hooked together along one edge, with stacked pyramid-like shapes inside. Though scientists were not able to analyze the crystals any closer, the crystals were visually identical.

Super Snowstorms: Winter Weather Traffic Patterns

Except for a band around the equator, all the world experiences varying degrees of the cold. It's a seasonal thing: The tilt of the Earth's axis produces the planet's seasons. One of the seasons is winter, alternating between the Southern and Northern Hemispheres, in June and December, respectively. Besides the poles, seasonal cold is especially felt in the middle latitudes just below the poles. There, colder air masses traveling from the polar regions are often fast moving and ferocious. In the United States, these air masses often produce snowstorms to beat all snowstorms, many with names that sound like all-star wrestlers.

One of the best-known bearers of such snowstorms are the *Nor'easters*, intense systems familiar to the populated eastern seacoast of the United States. The warm Gulf Stream current that runs along the coast sets the stage for the Nor'easters. As jet stream disturbances cause the air to rise over the Gulf

Stream, the temperature differences create a low-pressure system at the surface. Add to this equation abundant moisture from the ocean, temperature differences along the coast, and a strange system seems to pop up out of nowhere.

These suddenly occurring storms create a strong easterly flow of air to the north of the storm and a westerly flow behind it; in fact, they are named Nor'easters because the winds generally blow from the northeast. High winds rage and heavy rains drop along the coast. Inland, the rain can turn to ice and snow as the Nor'easter hits the cooler air usually in place at that time of year. Though such storms usually travel fast—they end when drier air from the west displaces the storm—they can still create havoc. A classic example occurred in 1993, when a superstorm along the eastern coast of the United States killed 270 people and caused approximately $1.6 billion in damage. In the northeast, people shoveled out from record snowfalls. Those east of the Great Lakes were hit even worse. These cities experienced tremendous *lake-effect* snows, forming as strong west-northwest winds in the wake of the Nor'easter pulled moisture off the lakes and dropped snow.

Alberta Clippers (or Saskatchewan Screamers, if the storms start farther east in Saskatchewan) do not always drop large amounts of snow like a Nor'easter, but they do increase the chance for cold air and high winds to move into the affected areas. As the name implies, the Alberta Clipper starts on the lee (eastern) side of the Canadian Rockies. The air flows rapidly down the mountains; friction from the air interacting with the mountains creates eddies and swirling winds, forming a low-pressure center at the surface. The combination of the fast-moving air from the steep pressure gradient, and the cold air pulled down from northern Canada, creates the fast-moving, cold, windy Alberta Clipper. Owing to the low moisture in the system, these "clippers" usually drop only a dusting of snow. But under the right conditions—especially when the winds grab moisture from the Great Lakes or even the Atlantic Ocean—those unlucky northeastern cities can receive hills of snow from the Alberta Clipper, only from a different direction.

• The Other Winter Storms •

Additional winter storms, mostly in the United States, come in under a flurry of different circumstances:

• The commonly occurring *Pineapple Express* brings warm air from around Hawaii, and rain to the West Coast, from the Pacific Northwest to California. If conditions are right in the winter, the warm air rises into the higher elevations of the Sierra Nevada or Cascade Range, sparking heavy snow as the air cools.

• The *Texas Panhandler* (Colorado Low, or Panhandle Hooker) forms as cold winter air flows down the Rockies, then collides and mixes around the flat Panhandle with warmer, moist air from the Gulf of Mexico. As the warmer air rides over the colder air from Canada, snow falls in the Plains and Midwest states, fed by the abundant moisture being supplied by the Gulf.

• Another storm that gains strength over eastern Tennessee, the *Chattanooga Choo-Choo* (CCC), forms when a jet stream disturbance causes a low pressure to form along a cold front trying to get over the Appalachian Mountains. The CCCs grab cold air from the Great Lakes, turning any moisture associated with the slow-moving storm into snow in the Tennessee and Ohio Valleys.

• The *Nantucket Effect* forms as a low-pressure system develops in the Atlantic Ocean, southeast of Nantucket, Massachusetts.

The effect of this seemingly simple low has, in the past, been devastating. The Nantucket Effect dropped more than 2 feet (0.61
meters) of snow throughout the northeast on January 6, 1996, apt-
ly named "the Blizzard of 1996." The effect is also known to occur
in warmer months, such as the May 3, 1994 "mini-hurricane," a
storm that produced easterly winds over 70 miles (113 kilometers) per hour and torrential rains along the northeast coast.

Thunderstorms: Symphonies in the Sky

Thunderstorms are pretty common, with more than 40,000
occurring throughout the world each day. There are dozens of

This nor'easter ran up the eastern coast of the United States in October 1996.
The view is typical for such a storm—a huge cloud system spinning counterclockwise, gathering moisture and speed from the warmer ocean waters. (photo
courtesy of the National Oceanic and Atmospheric Administration, taken by the
GOES-8 satellite)

ways in which thunderstorms are created, but by far, most form from the collisions between a warm front and cold front or by convection. The majority form with three simple, basic ingredients: moisture, cooler air with height, and something to "push" the moist air from near the ground.

Simply put, the rising warm air continually feeds the thunderstorm cloud cell. This creates billowy, thick cumulonimbus clouds, which, depending on circumstances, can be only a few miles in diameter or span hundreds of miles. Up-and-down movements of air are fed by rising warm air from the surface, or even the release of heat as water vapor turns to liquid (and liquid to ice) to form the clouds. As the storm matures, growing to heights above 20,000 feet (6,096 meters), a flow of cold air falls downward or to the ground, signaling the weakening of the storm. Depending on the type of thunderstorm, it can last from fifteen minutes to several hours, the downdrafts eventually cutting off the supply of warm air from the surface.

You've probably experienced one classic thunderstorm scenario in the early afternoon in spring or summer. The day is usually hot and sunny, with only an occasional cloud in the sky. Toward afternoon, the upper-level air becomes unstable as moist air near the surface heats up and rises, like steam in a Turkish bath. At first, the instability causes fleecy clouds, and if the conditions are ripe, they eventually build into tall, billowing clouds, usually resulting in the development of a thunderstorm by mid- to late afternoon. Most of these nonfrontal, convective storms travel by quickly, clearing the air of particles and often creating a rainbow after they pass, especially since the Sun is lower on the afternoon horizon. Convection initiates this type of thunderstorm, called *thermal* or *air-mass thunderstorms*.

The three basic kinds of thunderstorms—mostly air-mass

thunderstorms—include *single cell, multicell clusters,* and *multicell lines.* Single cells typically last twenty to thirty minutes and can produce some downbursts, heavy rain, hail, and sometimes even a weak tornado. Multicell-cluster storms are a group of cells moving as a single unit, with each cell at a different stage in the thunderstorm's life; they can produce moderate-size hail, flash floods, and weak tornadoes. This forces warmer surface air to rise, thus fueling the further development of other cells nearby. The multicell-line storms, or squall lines, are a band of clouds with a continuous, developed gusty wind (called a gust front) at the front of the line. Similar to a multicell cluster, they produce moderate-size hail, occasional flash floods, and weak tornadoes.

Storms that typically arise from a deep convection process often develop into large-scale storms, called *mesoscale convective systems* (MCSs), with the largest (and longest lasting) of them called the *mesoscale convective complexes* (MCCs). The storms are often preceded by isolated thunderstorms that merge in the late afternoons. Usually, by nighttime a low-level jet stream develops, feeding warm, moist air into the cluster. This builds up the complex, allowing the storm to last six hours or more, which is why these massive storms usually peak near midnight.

The MCCs are immense, covering an area nearly the size of several small states. If they move slowly enough, they can drop flash-flood-type rains. And they are long-lived, sometimes lasting twelve hours or more. MCCs develop all over the world, including cloud systems over the equatorial western Pacific Ocean; over the Indian subcontinent, where they contribute to the Asian monsoon cycle; over the land in Africa, mostly in Sahel; and over the mountains of western Mexico. Strangely enough, in moist, warm areas such as the equatorial rain forests of Africa, there are few MCCs.

These thunderstorms above several islands in the Atlantic Ocean show how high the storms reach into the upper atmosphere, producing amazing cloud shadows, as seen on space shuttle mission STS-48. (courtesy of NASA)

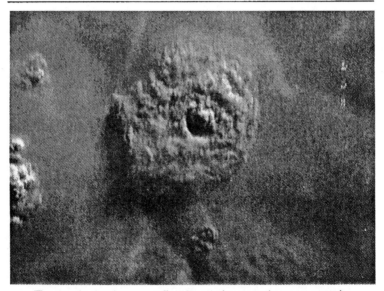

The top of a thunderstorm cell, with a circular eye in the center, over the Atlantic Ocean, as seen on space shuttle mission STS-45. (courtesy of NASA)

Scattered "air mass" thunderstorms over Zaire, Africa, October 1994.
(courtesy of NASA)

• The Bizarre That Fell from the Sky •

Be careful where you step. In some places around the world, things drop out of the sky that have nothing to do with the clouds. Here are a few "droppings from the sky," seen mostly during intense and severe thunderstorms:

• It may be a hard way to seed the ground, but in 1687, hail containing the seeds of ivy berries fell in England.

• In 1892, a thunderstorm in Germany resulted not only in rain but a rain of hundreds of freshwater mussels.

• Fresh frozen ducks inside hail were dropped during a 1933 storm in Worcester, Massachusetts.

* Small unripe peaches were dropped during a passing thunderstorm in 1961, on a portion of Shreveport, Louisiana.
* In 1995, a tornado swept through Moberly, Missouri. It dropped unopened soda cans from a Moberly bottling plant in Keokuk, Iowa, 150 miles (241 kilometers) away.
* Toads rained down on the Mexican town of Villa Angel Flores in 1997 after a tornado grabbed a cluster of the amphibians from a local body of water.

Inside a Thunderstorm: Up Close and Personal

Since the dawn of aviation, several people have fallen through a thunderstorm and lived to recount the ordeal. One of the most well known is Lieutenant Colonel William H. Rankin, who, in 1959, was forced to eject over Norfolk, Virginia, from his F8U supersonic jet at 47,000 feet (14,326 meters). Not only did he have to endure the effects of rapid decompression as he fell through the thinner part of the atmosphere, he had to contend with no oxygen supply and frostbite from the tremendous cold (−70°F or −57°C). When he hit 10,000 feet (3,048 meters), the oxygen was more plentiful—but his adventure was just beginning: Like the hapless occupant of an amusement park ride gone wrong, Rankin was treated to a harrowing forty-minute ride as he plunged right into the maw of a severe thunderstorm.

Just what is going on inside a thunderstorm? Most of us will never know. But for William Rankin, it was an experience he would never forget: "It hit me like a tidal wave of air, a massive blast, as though forged under tremendous compression, aimed and fired at me with the savagery of a cannon. I was pushed up, pushed down, stretched, slammed, and pounded. . . . I was jarred from head to toe. Every bone in my body must have rattled, and I went soaring up and up as though there would be no end. As I came down again, I saw that I was in an angry ocean of boiling clouds, blacks, grays, and whites spilling over each other, into each other, digesting each other." Rankin was being carried up and down within the thunderstorm's updrafts and downdrafts, the normal turbulence within the huge thunderstorm complexes.

Rankin's ordeal did not stop there. He was also treated to the usual accompaniment in severe thunderstorms: lightning, hail, and rain. He described the thunderclaps as unbearably loud physical experiences and further explained that the lightning seemed to be everywhere. "I saw it in every shape imaginable. But very close, it appeared mainly as a huge bluish sheet, several feet thick, sometimes sticking close to me in pairs, like the blades of a scissors, and I had the distinct feeling that I was being sliced in two." The hail gave him welts where it struck his drenched clothes, and the rain inside the cloud was just as fierce. "Sometimes the rain was so dense, and came in such swift drenching sheets, I thought I would drown in midair. 'How silly,' I thought, 'they're going to find you hanging from some tree, in your parachute harness, your lungs filled with water, wondering how on Earth you drowned.'"

Curiously and miraculously, Rankin landed in a tree—although he slammed into the tree's trunk as he landed—safely making it through the usual downdraft winds at the base of a thunderstorm. Even more amazing, he was able to walk and flag down help in Rich Square, North Carolina, 65 miles (105 kilometers) from where he bailed out.

• To Boldly Go . . . •

It takes a special type of person, and a special type of plane, to fly into a thunderstorm. The King Air research aircraft, a twin turbo-prop plane, collects data on everything from cloud physics and atmospheric chemicals to mesoscale dynamics of the clouds. The EC-130Q Hercules, operated by the National Center for Atmospheric Research (NCAR), flies through clouds for research on electrical fields and the formation of precipitation. There is even NCAR's sailplane, the L188C Electra, which collects data on minute changes in the forming thunderstorms, lifted by the developing storm's upward-moving air.

But there is only one plane so far in the world that flies through a mature thunderstorm: the T-28, a former trainer plane built for the military in 1949. Although the designers of the T-28 did not intend it for storm use, it's a sturdy and rugged plane that has been used to study lightning, hail, and electrification data since 1969. The plane is specially equipped to face the rigors of the storm, having a windshield 3/4 inch (1.9 centimeters) thick and steel bracing over the pilot's canopy for protection from damaging hail. In addition, the tail is armored, as are the wings' leading edges.

5

Weird Lightning Phenomena

I n the sports world, numerous games have been postponed or canceled because of weather, especially thunderstorms. Tennis championships and baseball games have been rained out, and outdoor football games have become muddy slideball games from the downpours of thunderstorms. The one sport that seems to invite danger from lightning is golf. Although golfers are not struck by lightning significantly more than other outdoor sports participants, it has been reported that of all persons who have been struck while taking shelter under a tree, a third have been golfers. In 1982, the National Weather Service (NWS) began attempting to reduce the dangers to professional golf players at the Tulsa, Oklahoma, tournament. Using storm spotters and radar, the NWS tracked potential storms in the storm-ridden state, displaying a green flag for good weather, yellow if a storm with lightning is approaching within two hours, and a red flag if lightning is likely in thirty minutes. One overall rule applies to professionals and spectators alike: If lightning threatens, get inside and don't seek shelter under trees.

Lightning: Zeus Trying to Get Our Attention

Lightning is one of the most fascinating aspects of our surrounding atmosphere. This discharge of electrical energy resembles a giant spark. The visible section is called a lightning bolt, or streak, forked, jagged, or zigzagged lightning, and forms at about 15,000 to 20,000 feet (4,572 to 6,096 meters) high, usually in thunderstorm clouds.

Just how lightning develops is still not totally clear, but some scientists believe that ice crystals in the upper part of a storm cloud are generally positively charged, whereas water droplets at the bottom are generally negatively charged. As the upward and downward movements of air carry the two respective charges throughout the clouds, the positive charges head for the top of the cloud and the negative charges go below. You have probably experienced such a phenomenon: When you shuffle your feet across a carpet (especially when the weather is cold and dry), you can attract positive charges from the carpet to your body. Touch something metal, or even the person standing next to you, and you can feel the shock of static electricity as the negative charges within the object or person meet the positive charges.

In the case of the cloud, the built-up charges have to go somewhere, and such discharges occur as cloud-to-cloud (discharge between two thunderstorms), intercloud (discharge between two clouds), cloud-to-ground, and in-cloud (intracloud or sheet) lightning. Yet the discharge that captures our attention the most is cloud-to-ground lightning. The negative charge near the storm cloud's bottom attracts a "shadow" of a positive charge on the ground below. When the voltage difference between the charges is great enough, the negative charge moves toward the positive charge in an invisible pattern called a step leader; the descending leader attracts the ground's positive charge upward, usually through a tall object such as a tree or lightning rod. When the positive and negative charges meet, the lightning bolt results, the bright tongue of light heading from the ground to the cloud (the opposite of what most people believe), sustained by a return positive charge back to the cloud. It takes only a fraction of a second for the charges to cancel each other out in a flash of lightning. The next step is more audible: The sudden expansion of air due to heating produces a sound wave that propagates from the source, reaching our ears as a rumble of thunder.

Overall, the standard estimate for the occurrence of lightning around the world is 40 to 50 flashes per second (the original number was 100 flashes per second, but with new satellite readings, the estimate has gone down). In the lower forty-eight states, cloud-to-ground lightning averages about 20 million to 30 million strikes per year, with the majority occurring in Florida between Tampa and Orlando, thanks to the high moisture and temperatures, especially in the summer. It's also a matter of perspective too: The top of the Empire State Building is hit about 500 times a year—and was once struck fifteen times in as many minutes.

More than half of all lightning bolts strike several spots on the ground. And on the average, there are five to ten times as many intra- and intercloud lightning flashes as cloud-to-ground. A big bolt of lightning may seem to be hundreds of feet wide when it hits the ground, but actually it measures only about an inch (2.54 centimeters)—just thicker than a pencil; the actual flash is three times hotter than the surface of the Sun; and the peak current can be more than 300,000 amperes. Lightning has been known to strike people, animals, trees, planes, and structures. Those standing close to a lightning strike often smell the hot blast changing regular oxygen, with two oxygen atoms, to the three oxygen atoms of ozone.

What You See: "Different" Lightning

There are no "different types" of lightning—they only *look* different. One in particular is merely a fantasy: *heat lightning*, so called because it appears to brew from silent, heat-generated storms that often occur during the hotter days of summer. In reality, the heat lightning seen from far away is lightning over the horizon reflecting off high cirrus clouds or haze. Closer heat lightning is often intracloud flashes, in most cases, more than 5 miles (8 kilometers) away. And since thunder is rarely heard at such distances, they appear to move silently across the sky.

Anvil crawlers appear to crawl across the sky and are probably cloud-to-ground lightning, branching upward and outward like a tree at the top and sides of the largest of thunderstorms. Rare *bead lightning* resembles a string of bright beads, when certain sections of a bolt remain lit for longer than usual. Wind blows the return stroke of a lightning bolt sideways, creating streamers called *ribbon lightning*. When lightning looks as if it is flickering on and off, this could be caused by *stream-*

ers, minute discharges within thunderclouds tha

lightning flash, causing it to flash over and over.

• Lightning Energy Equivalents •

How does the average bolt of electricity match up to o

tastic explosions of energy? Here are some equivalents

amount of energy released by each event; these are all norr

to 1 million units of work (scientists call them ergs), so the ɛ

can be compared:

100-watt lightbulb left on for a week	630
One-ton car going 25 miles (40 kilometers) per hour	630,000
Bolt of lightning	630,000,000
Explosion of a 1-kiloton atomic bomb	630,000,000
Eruption of Mt. St. Helens in 1980	630,000,000,000,000
Annual U.S. energy consumption	630,000,000,000,000,000

(Source: National Lightning Safety Institute)

• For a link to the National Lightning Safety Institute, try:

http://www.lightningsafety.com/

Who Gets Hit How: Chances of Lightning Strikes

On January 6, 1993, during a freak winter lightning storm in
Eynon, Pennsylvania, lighting struck a yard, breaking a water

Lightning is a buildup of electric energy in the atmosphere released in the form of a lightning bolt, as seen in this photo of a cloud-to-ground lightning bolt. (photo courtesy of Grant Goodge, National Climatic Data Center)

main and sending a roof flying 80 feet (24 meters) away. The flash continued through a house cellar, damaging the foundation; it rolled out of the front of the house, blowing out a garage wall.

The peninsula of Florida, called the lightning capital of the world, has, on average, the highest number of lightning strikes per year (although the "honor" was temporarily passed to the state of Missouri in the summer of 1993). In east central Florida alone, there are approximately 90 to 100 thunderstorm days per year, primarily in the summer months. In 1998, lightning storms did more damage than usual, igniting parched vegetation and causing massive wildfires to burn out of control.

On average, lightning causes more casualties annually in the United States than any other storm-related phenomenon, except floods. In fact, the statistics—75 to 150 people killed each year—may be underreported as more are misdiagnosed as burns or cardiac problems. Here are some other statistics: Men are struck by lightning eighty-four percent more than females; the majority of the strikes occur in July, mostly on Sundays, Wednesdays, and Saturdays. The five states that have the most deaths are, in order, Florida, Michigan, Texas, New York, and Tennessee, whereas the injuries, in order, occur mostly in Florida, Michigan, Pennsylvania, North Carolina, and New York—for no real explainable reason.

Of course, the United States is not the only place where lightning kills. One report notes the mean number of lightning deaths per million people around the world each year. 0.2 per million in the United Kingdom, 0.6 per million in the United States, 1.5 per million in South Africa, and a "whopping" 1.7 per million killed in Singapore.

• When to Head for Cover •

Though it is harder to do in mountains (thunder echoes off the hills), or to distinguish between multitudes of lightning flashes and thunder during extremely violent storms, there is a rough way to determine your distance from most thunderstorms. Because light waves travel faster than sound, we see lightning first, then hear the thunder. Thunder travels about 1 mile in 5 seconds, or 1 kilometer in 3 seconds.

To calculate the distance to the storm, count the seconds between seeing the lightning and hearing the thunder; then divide by 5 for miles, 3 for kilometers. For example, if you count to 10 seconds, the storm is 2 miles (3.2 kilometers) away. If the thunder follows the flash almost immediately, the storm is probably under a quarter of a mile away. If you hear a flash and a crack immediately, it could be within 100 yards (91 meters) away! If your answer is any more than 20 miles (32 kilometers), however, it is doubtful your calculator is working. In general, you can't hear the thunder at such long distances because the atmosphere scatters and absorbs the sound waves.

Why watch out for lightning? Because it can strike in many common places. According to the National Oceanic and Atmospheric Administration, during a span of thirty-five years, from 1959 to 1994, the situations in which people were struck by lightning had a pattern—and here is the breakdown:

40 percent - unspecified

27 percent - in open fields and recreation areas (but not associated with golf)

14 percent - under trees (but not associated with golf)

8 percent - water related (boating, fishing, swimming, etc.)

5 percent - golfing, including hiding under trees while golfing

3 percent - heavy equipment and machinery related

2.4 percent - telephone related

0.6 percent - radio, transmitter, and antenna related

Noisy Reaction: Booming Thunder

Thunder is a natural reaction from a flash of lightning, the sound produced by the explosive expansion of air in the lightning channel, heated on the order of three times the temperature of the sun's surface. At first, the resulting supersonic shock wave moves faster than the speed of sound, then gradually slows a few hundred yards away, turning into a sound wave as it moves away from the channel. Although the lightning bolt and thunder occur at about the same time, light travels very fast—at about 186,000 miles (299,274 kilometers) per second. Thus, the flash is seen first, and the thunder clap, traveling at a snail's pace of one-fifth of a mile in the same time, is heard a short time later.

There are various thunder booms, too. The closer to the lightning flash, the more high pitched and "crackle-sounding" the thunder; farther away, the pitch is lower and a more booming or rumbling sound is heard. Prolonged thunder rumbles occur in valleys as the sound bounces off the surrounding hills, or when the lightning channel is crooked and jagged, causing the sound to arrive at different times and directions. If you're really close to the lightning strike, within 300 feet (91 meters), you'll hear a loud, startling, high-pitched bang, actually the shock wave that has yet to turn into the normal thunderclap.

And there is one more sound effect of lightning you can hear if you're listening to an AM station on the radio as a storm approaches. The static bursts you hear are *sferics*—the radio waves produced by the lightning discharges.

Ball and Other Lightning: Great Balls of Fire

Patricia's grandmother was once standing back from the counter after washing dishes at her farmhouse. It was the summer of 1944, and a dark thunderstorm was rolling through the area. Suddenly, she saw a ball of lightning no larger than a softball flash through the open window near the sink. As she watched, the ball seemed to spin around the room, zigzag a few times, then vanish into the living room. She never saw where the ball finally popped out, for there were no scorched holes anywhere as evidence of its escape. But several pans in her dishwater were fused together, a reminder of the amazing power of this phenomenon.

Ball lightning, sometimes called a fireball (although this term is often used in astronomy to describe a bright meteor), is thought to be a cloud-to-ground lightning that manifests itself in the shape of a ball. Although rare, it has been reported everywhere: On June 8, 1977, a yellow-green ball of lightning the size of a bus bounced down a Welsh hillside for about three seconds. In Turkey, in June 1988, several observers from a fifth-floor office witnessed a bright circular flash of blue-purple light appear and vanish within two seconds.

Amazingly, those who have seen ball lightning report not only the swiftness of the arrival and departure of the ball, but also a bad smell. And in a recent study, scientists gathered thousands of reports of apparent ball lightning events. In many of the cases, lightning balls apparently bored holes in windows, leaving perfect round discs where the lightning entered and exited panes of glass. Other ball lightning left no such holes, the lighting passing right through as if the glass didn't exist.

These glowing, skipping balls of light are some of the most transient phenomena associated with lightning storms. No one really understands what causes ball lightning, but many scientists believe the glowing balls are made of plasma created as an offshoot of the lightning (while still others do not believe in the phenomenon at all). In addition, some reports point to possible ball lightning inside funnel clouds, which may account for reports of luminous tornadoes, those that appear illuminated from the inside.

Another even rarer form of lightning is called *Saint Elmo's fire*. This phenomenon, named after Saint Erasmus, the patron saint of sailors who popularized his name to Saint Elmo, resembles a mass of sparks high above the ground near a thunderstorm. The crest of sparks, often glowing blue, forms when the buildup of opposite electric charges is not strong enough to create a flash and the lightning bolt fizzles out. First spotted at the top of ships' masts (where they were considered a good omen), Saint Elmo's fires are typically seen around tall metal grounded objects, such as lightning rods, aircraft wings, and chimney tops.

• *For more information on ball lightning, including excerpts from the 1990 Journal of Meteorology, try linking to:*
http://www.knowledge.co.uk/frontiers/sf071/sf071g17.htm

• Bolt from the Blue? •

In 1995, a lightning bolt hit a playing field near Miami, Florida, injuring ten children and a coach. So why didn't the coach get the children inside when he saw the storm? In this case, the sky was clear, the closest line of clouds at least 10 miles (16 kilometers) away to the northwest.

This was a classic case of a "bolt from the blue," or *anvil light-ning*, in which a faraway storm system delivers a stray lightning bolt to a distant area with a clear sky. In most cases, the flash emanates from the rising cloud (caused by the upward movement of air, or updraft) of the thunderstorm. It then travels sometimes more than 25 miles (40 kilometers) from the originating storm, angling downward to contact the ground.

• Fossilized Lightning •

One of the more interesting finds in the desert are objects called *fulgurites*, from the Greek *fulgur* ("lightning"), directly resulting from the great thunderstorms that roll through the dry areas. These finger-shaped chunks of natural glass are tough to find, but they do have a striking history, so to speak. They form when 40,000°F (22,204°C) lightning strikes certain types of dry sandy soil, essentially freezing the path of the lightning as it strikes and fuses the grains of sand. Fulgurites have been found in small chunks or larger branches. In Michigan, a fulgurite reaching about 15 feet (4.6 meters) in length is the largest found so far, with three branches, one measuring over 16 feet (4.9 meters), one over 14 feet (4.3 meters), and the last about 8 feet (2.4 meters) in length. And in the Great Sands National Park in Colorado, there are signs posted to hikers traveling the dunes, warning them to take cover during thunderstorms—and "not to become fulgurites."

Sprites, Jets, and Elves: Wizardry in the Sky

For more than a century, people have seen strange nighttime flashes over the tops of thunderstorms. A 1903 paper talked about "rocket lightning" distinguished by "a luminous tail . . . shooting straight up . . . rather faster than a rocket." Reports from Africa in 1937 talked about long, reddish streamers. In the 1950s, English scientific papers described what seemed to be flames shooting above thunderstorms that hugged the horizon. But no one really believed that the sights were anything but the reflections of lightning from the clouds below, mere illusions.

Technology ultimately changed all that. In 1989, several University of Minnesota scientists who were testing a low-light video camera for a high-altitude scientific rocket shot captured two huge pillars of light above a distant thunderstorm. University of Minnesota scientists John Winckler, Robert Franz, and Robert Nemzek recorded giant twin pillars of light reaching more than 18 miles (29 kilometers) above a distant thunderstorm. Scientists now whimsically call these brief, colored flashes of light *sprites, blue jets,* and *elves*—also once called cloud-to-sky lightning.

Red sprites (sometimes with bluish tendrils) flash high above thunderstorm clouds, often as high as 60 miles (95 kilometers). They come in all shapes and sizes, from giant red blobs, picket fences, thick tentacled octopi or jellyfish to branching carrots. The flashes were named sprites after the creatures in Shakespeare's *The Tempest,* in part because of their transient, ephemeral nature. But unlike the Bard's characters, these sprites are very real indeed.

Some storms produce the sprites frequently, but the sprites are still considered rare; they can occur in clusters or singly. Scientists believe that charges within the thundercloud are responsible. The upper, positively charged parts of the cloud often discharge to the ground, producing a super-intense lightning flash. About one in twenty of such cloud-to-ground lightning bolts are so energetic that they spawn the sprites; in this case, not all the charge goes into the flash. Similar to the zap you feel when rubbing your feet over a rug on a cold, dry day and touching a piece of metal, the extra electrostatic charges build up in the cloud. If the lightning discharge is large enough, it "ignites" the extra charge above, creating a sprite.

One of the best places to pick out and image these elusive weather spirits is on mountaintops with a great view of a flat plain below. Scientists at Langmuir Laboratory in New Mexico found a prime spot, on the lab's own 11,000-foot (3,353-meter) peak, which provides clear views of thunderstorms in Kansas, Colorado, Texas, and northwestern New Mexico. Their watch proved productive: Detailed images of the red blobs of sprites consist of thousands of fiery *streamers*, each only a few feet wide. The tendrils branch and turn on and off but last less than 17 milliseconds.

There's more: Mysterious blue jets can be seen with the naked eye from the ground, but they are rare and difficult to spot. The first blue jet was imaged above an intense Arkansas storm in 1994, by scientists Davis D. Sentman and Eugene M. Wescott, and it appears that the lights are best seen in intense hailstorms. The blue beams of light jet upward from the top of the thunderclouds at about 75 miles (121 kilometers) per second, reaching heights two or three times higher than the clouds, then they blink out of existence.

Related high-altitude flashes are elves. These giant expanding disks of light, some more than 250 miles (402 kilometers) in diameter, are also created by exceptionally strong conventional lightning. Although it sounds like something out of *Star Trek*, scientists believe electromagnetic pulses, essentially intense bursts of radio static, may be the reason for the elves. When the bursts given off by these intense discharges reach a critical height, the electrons in the field strike air molecules, knocking them into an excited state that releases light. Elves can occur with sprites. Elves tend to form first, but they vanish more quickly, in less than one-thousandth of a second, too fast for the human eye to see.

• For more information on sprites, bluejets, and elves, try linking to:

http://elf.gi.alaska.edu/

and FMA Research, Inc., Fort Collins, Colorado:

http://www.fma-research.com/

• Spotting the Sprite •

How can you see any of these strange atmospheric appearances? The easiest flashes seen from the ground are the red sprites. They are by far the most common of these mesospheric creatures, and we know where they "live." So a plan for some serious "sprite hunting" is relatively easy to develop.

Look for intense thunderstorms at night, usually clusters of storms at least 100 miles (161 kilometers) long. To improve your chances, watch your local TV channel or get on a weather radar site on the Internet. Look for thunderstorm clusters that are at least 93 miles (150 kilometers) long on one side. It's also best to watch the thunderstorms from about 30 to 60 miles (48 to 97 kilometers) away so you can easily see the tops of the clouds, so note where the thunderstorms are located.

Now find a good view of the horizon, preferably away from city lights, and on a night without moonlight. After your eyes adjust to the dark, look toward the top of the cloud about four to five times the height of the cloud tops, not the storm itself. Using your hand, a piece of cardboard, or even a building, block out the lightning flashes below. Then be extremely patient and don't blink, as they last for only one-hundredth to one-tenth of a second. While some thunderstorms never produce a sprite, others can build up enough charge, creating a sprite every one to ten minutes. What will you see? Most people describe it as an aurora that turns on and off in a second, including the different colors of red, green, orange, or white. The varying colors are because your eyes have a problem interpreting true colors in low-light conditions.

The best places in North America for sprite watching? Probably above the northern High Plains and upper Midwest, in a broad belt from Colorado to North Dakota, over to Minnesota and down into Texas. But they do occur above big storms worldwide, and have been spotted from aircraft and the space shuttle above Panama, Peru, Africa, Australia, and Indonesia, to name a few places.

6

Falling Winds

On top of Mount Washington in New Hampshire's White Mountains, the highest wind speed ever officially recorded occurred on April 12, 1934, measured at 233 miles (375 kilometers) per hour. In a strange turn of events, and during an El Niño year, a typhoon called Paka ripped past Guam on December 16, 1997, the instruments recording some of the highest winds ever recorded on Earth: 235 miles (378 kilometers) per hour. But scientists determined that the instruments used to measure this wind were not calibrated correctly, and thus Mount Washington still holds the record. But, of course, they have nothing on the winds on Saturn—the highest clocked at 1,641 feet (500 meters) per second at the planet's equator, or about 1,120 miles (1,802 kilometers) per hour.

Winds: Whipping Up the Air

The winds can carry a startling punch. (This refers not to the usual breezes on a bright sunny day but rather to the winds spawned primarily by severe storms.) It is almost as if we were the victims of continual fights between warm fronts and cold fronts, or the convective cells that produce severe storms. Gravity doesn't help, either, pulling the winds down to the surface. The ancients even named them: Aeolus was the god of the wind, with the north wind called Boreas, the south wind Auster, the east wind Eurus, and the west wind Zephyrus.

To understand how wind develops, consider something we have all done before: Blow up a balloon and release it. As we breathe into the balloon, we are forcing the air to stretch the rubber thinner and create more volume inside. We have, in effect, created our own high-pressure system. When the end of the balloon is released, the high-pressure air inside moves toward the lower pressure outside, equalizing the pressure and creating what we call wind.

Though balloons are *not* responsible for our planet's windy weather, this scenario is similar: The pressure difference between dry and humid air masses—or high- and low-pressure systems—moves the higher-pressure air toward the lower-pressure air, creating the winds that we feel. And the greater the pressure differential, the stronger the winds.

• Winds and Hot Air Balloons •

Richard Branson and Per Linstrand were the first two balloonists to cross the Atlantic Ocean in a hot-air balloon, in 1987, and the first to cross the Pacific in 1991. In recent years, numerous attempts—by not only these two balloonists but other enthusiasts—have been made to be the first to fly nonstop around the globe by balloon. This was done in March, 1999, by balloonists Bertrand Piccard and Brian Jones.

To accomplish this trip within the reasonable time frame of just over two weeks, the balloons are designed to travel at high altitudes, using the jet streams for extra high-speed "propulsion." The attempts have been made in the Northern Hemisphere's winter, when the jet streams quicken up to 300 miles (480 kilometers) per hour; the downside is that the balloons must be at the right altitude, and in the right location, to catch these winds.

Every team of balloonists includes meteorologists who track the jet streams and progress of the balloon, advising the pilots where, when, and how to catch a faster ride. Unfortunately, the jet streams often flow over countries that don't want balloonists to fly over. In such cases, a substitute route must be found—or if the alternate path takes the balloonist too far off course, the flight will be aborted.

* For more information on hot air balloons, try
the World Wide Web balloon pages:
http://www.euronet.nl/users/jdewilde/index.html

Downdrafts: Downbursts from Above

We've seen downbursts and microbursts many times in our travels. One memorable *downburst*, a rapid downward motion of air, slapped down the tail of a plane taking off from the

Denver airport, causing it almost to bounce and scrape across the runway. A more confined burst of downward-falling air, called a *microburst* also caught up with us in Colorado Junction one year. A thunderous storm rolled out of the west, bringing with it plenty of lightning, thunder, and winds. We stood outside of our first-floor hotel room in the middle of the afternoon as the clouds turned the sky to night. The winds howled, bending nearby trees in half and rocking a nearby truck, its wheels actually lifting off the ground as if it were shaken by some unseen hand. The microburst passed us in less than 5 minutes, but not before it did its damage. Several bushes near the outdoor swimming pool were torn from the ground, and a tree fell on the lot next door. Later reports mentioned the flipping of a small plane at the local airport near the hotel, localized power outages, and more downed trees. The microburst winds measured greater than 70 miles (112 kilometers) per hour.

Strong winds can vary greatly for a number of reasons, most having to do with thunderstorms. The winds of a mature thunderstorm not only rise but fall, the result of the powerful up-and-down movement of the air associated with the upper levels of such storms. The best-known downward motion of the air that can actually reach the ground is called a *downdraft*, the downward motion of air in the storm cloud.

While the upward motion of air (*updrafts*) in a storm cloud can reach tens of miles per hour, downdrafts can travel at equal or more intense speeds. A multitude of factors can cause downdrafts, including the downward drag from heavy masses of rain and hail, the precipitation within the cloud cooling the air as it evaporates (making it heavier), plus gravity's pull on the cooler, heavier air. You can often feel a mild downdraft in the summertime just before a rainstorm—a cool rush of air that spreads out ahead of the rain. One minute, you are warm; the next, the temperature drops, sometimes as much as 20 degrees, and you feel wind called the *gust front*, the leading edge of the downdraft.

The more powerful downdrafts produce downbursts, which are so intense the falling winds seem to smash into the surface, with gusts often reaching into the hurricane-force range of higher than 74 miles (119 kilometers) per hour. First named by Theodore Fujita in 1976, though even Leonardo da Vinci drew them in his famous notebooks, downbursts are concentrated, severe downdrafts that cause an outward burst of damaging winds on the ground. Downbursts are relatively small-scale phenomena, generally less than 6.2 miles (10 kilometers) in diameter, and are often confused for tornadoes. They are further divided into two groups: *macrobursts*, which cover larger areas, with outflow diameters greater than 2.5 miles (4 kilometers), and microbursts, or outflows of air with lesser diameters, often covering only several hundred yards in width.

Microbursts also come in three types: In the humid east, *"wet" microbursts* are most common; in the drier west, *"dry" microbursts* are more prevalent; and finally, in Texas and other states eastward along the Gulf of Mexico, there are intermediate wet-and-dry microbursts. Wet microbursts occur most often with heavy rain. Here again, evaporation causes the production of strong winds, but rain also reaches the ground. The result is a burst of wind and rain hitting the ground with such force that they seem to spread outward and upward in a distinctive curl—like a wave hitting rocks along a shore. Wet microbursts also create a *rain foot*, a distortion at the edge of the downburst cloud area. Dry microbursts occur when a column of rain falls into a layer of dry air beneath the cloud, creating *virga* conditions (in which the rain evaporates before it reaches the ground). When this happens, the rain cools the area, adding to the weight, which causes the air to fall. The result: a powerful gust of wind-raising dust and dirt called a *dust foot*. Not all microbursts reach the ground, where the air is warmer. This warmth tends to rise, counteracting the downdraft.

• *For more information on microbursts, try linking to:*

http://www.nssl.noaa.gov/~doswell/microbursts/Handbook.html

This dramatic dry microburst occurred in the southwestern United States and was triggered by a downdraft within a major thunderstorm. (photo courtesy of the National Center for Atmospheric Research, Boulder, Colorado)

• On a Smaller Scale •

Lesser down-flowing winds associated with thunderstorms can still pack a punch. *Bow echoes* are relatively new discoveries associated with thunderstorms. Meteorologists believe that if they can spot a line of thunderstorms developing a bow shape on Doppler radar, they may be able to predict the formation of these severe storms. The bow echoes of severe thunderstorms can create intense problems, including winds at 100 miles (161 kilometers) an hour. The bow echoes form as rain falling from squall-line thunderstorms creates small vortices, or eddies, in the wind flow behind the towering storms. These eddies cause the storm tower to tilt backward and bow, causing titanic mid-level winds to rush into the line from the rear. The rain then drags these rapid winds to the ground. They reach tremendous speeds, often damaging vegetation and structures on the ground.

A *roll*, or *shelf cloud*, is another, smaller scale result of a downdraft, visible without radar. These threatening, wedge-shaped clouds often precede a severe thunderstorm. They form when a violent, rain-cooled downdraft within the thunderstorm advances into warmer air, forming the cloud wedge, squeezing clouds in the middle of the thunderstorm. In general, they are often accompanied by extreme wind shear and turbulence, with winds up to 60 miles (97 kilometers) per hour at the leading edge of the storm.

Derechoes: Building Large-Scale Downdrafts

Our good friends have a cottage in the Thousand Islands, New York, along the island-dotted St. Lawrence River that runs from Lake Ontario to the Atlantic Ocean. In the early morning hours of July 15, 1995, a major storm was rolling through, one of the worst our friends had ever seen. Trees fell nearby and the power went out. For about a half hour, lightning was everywhere and the winds caused the rain to fall horizontally. The storm, later classified as a *derecho,* had started from a cluster of thunderstorms over Lake Superior. It intensified near Kingston, Ontario, rolled through the Adirondack Mountains, Massachusetts, and the northeastern tip of Connecticut, stopping only when it reached Rhode Island and mixed with the ocean waters. The winds were fierce, and before their anemometers broke, meteorologists at the airport in Watertown, New York, recorded wind speeds of 128 feet (39 meters) per second. Nearly 90 percent of the trees in the Five Points Wilderness area of the Adirondack Park were either damaged or destroyed. Overall, several million acres of trees were destroyed. Tragically, five fatalities were reported.

Derechoes may sound exotic, but in reality, they are another specialized form of a downdraft. They are the especially windy squall lines of thunderstorms forming in very humid weather with strong winds in the middle layers of the atmosphere's thick blanket. A cold outflow—originating from dry air a few miles above the ground that has been cooled as precipitation falls into it and evaporates—flows from the storm clouds; this cool pool of air meets with the warmer, humid air near the ground, the meeting creating a specific exchange of up and down drafts. The cooler, heavier air has its own momentum, plus the momentum of the fast, upper-air winds with it. A line of strong winds blasts down from the storm.

Classic warm-season derechoes (such as the July 1995 storm) most often form in late spring or summer when convection causes unstable conditions in the atmosphere. In addition, they are often associated with northwest winds, pushing the squall lines in a southeast-easterly direction. They are almost common across the Midwest, and there is even thought to be a derecho corridor extending from southern Minnesota to northern Ohio, in which northwest winds are prominent.

• Derechoes Dead Ahead •

To qualify as a derecho, the storm winds have to be stronger than 58 miles (93 kilometers) per hour, while the area affected must be at least 280 miles (451 kilometers) long. Thus the derecho is often just as damaging as a localized tornado, or more so.

These storms bring one downburst after another, with winds that can reach up to 100 miles (161 kilometers) per hour. The storms' widespread wind damage is notorious, and the storms move rapidly, traveling an average of 118 feet (36 meters) per second, either cutting through a swath of land or in a patchwork of individual swaths. Not that derechoes are anything new: The name *derecho* was coined in the 1880s by the director of the Iowa Weather Service, Gustavus Hinrichs. In Spanish, *derecho* means "straight ahead" or "direct." Thus, these storms are different from the tornado, whose name derives from the Spanish word for "turn."

Heat Bursts: Nature's Convection Ovens

It was after midnight on June 15, 1960, on the northwest side of Lake Whitney in Texas. The skies were clear and the temperature hovered above 80°F (26.7°C). Without warning, a wind began to howl and the temperature began to rise. For the unlucky Texans nearby, it felt like being in front of a blowtorch. The temperature rose to nearly 140°F (60°C) in only a few minutes, and the wind was clocked at 80 to 100 miles (129 to 161 kilometers) per hour. A cotton field was completely scorched, car radiators boiled over, and roofs were blown off. People wrapped themselves in wet sheets and towels for protection from the scorching air. Lake Whitney was in the midst of a *heat burst*, made even more terrifying by its nocturnal visitation. For nearly three hours the storm raged, and people thought the end of the world had come.

How did a seemingly clear sky produce such an avalanche of heat? Heat bursts are a kind of high-temperature downburst. The Texas storm had all the basic characteristics: strong, gusty winds, a dramatic rise in temperatures, disintegrating thunderstorms, and a fall in the atmospheric pressure and dew point.

A heat burst can develop anywhere around unstable, dying thunderstorms, but they are best known in the United States in Oklahoma and Texas, where thunderstorms are most prolific. Typically, the hot, humid air feeds a thunderstorm during the daytime, while under the storm, cooler, dry air prevails. As the sun sets, the "fuel" for the storm is cut off, and the storm dissipates. The rain left over at the top of the clouds falls into the lower—hence cooler and drier—part of the clouds, causing the rain to evaporate. This evaporation cools the air further, rendering it very heavy. It falls to the ground, but unlike the usual downburst, this air warms dramatically as the higher air pressure closer to the Earth's surface compresses it. Warming, it becomes somewhat lighter; the downward movement of the air slows but still does not stop, eventually slamming into the ground and spreading out as a hot, dry burst of wind.

Usually a heat burst is mild, but if the collapsing storm is large, and the air is allowed to warm up for a longer period of time, watch out. At Lake Whitney, the air continued to move downward even after the thunderstorm rains ended. And as the air fell, it warmed up 5.5 degrees for every 1,000 feet (305 meters) of the fall.

Wildfires: When We Make the Winds

Dry conditions across the world can spark a rash of wildfires, each one with the potential to become its own smaller version of a weather-making machine. After all, spinning columns of air need heat sources, and the hot fire, sometimes hotter than 800°F (427°C) if there are no strong surface winds, can produce vigorous updrafts. The Santa Ana winds—warm, dry downslope winds from the mountains—fanned the fires of the Old Topanga firestorm in southern California in November 1993, which spread to 200 acres in ten minutes. The resulting updraft reached 6 miles (9.7 kilometers) into the air. In May through June of 1998, wildfires also burned thousands of acres of vegetation in Florida to the ground. The sea breezes kicked up the embers, and extreme hot spots fed on themselves.

A wildfire (or firestorm) modifies the wind, or actually produces its own winds, continually spreading the fire. The winds can also create a cloud that grows upward like a thunderstorm, complete with thunder, lightning, and heavy rains; these rains, by the way, do little to put out the raging fire. In addition, the fiery winds can swirl like tornadoes formed in supercell thunderstorms. Although these wildfire-spawned spinning winds rarely last more than a few minutes, they can move chaotically,

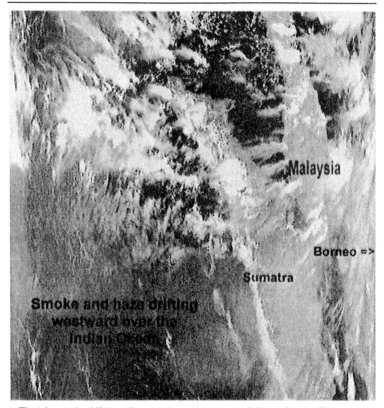

This photo of wildfires in Borneo shows the extent of their influence: The smoke and haze they produce reach hundreds of miles away. (photo courtesy of the National Oceanic and Atmospheric Administration, from the *NOAA-14* satellite)

sparking new fires as they wend their way around the blaze.

A *blowup* is a wildfire that grows more intense from a strong wind, like blowing on a campfire to increase the flames. The extra winds may come from a low-level jet stream or may even be self-generated by the wildfire. The forests of the United States, Canada, and Australia have endured classic blowups. One famous one was the Peshtigo fire of northeastern Wisconsin, which occurred at the same time as the great Chicago fire.

7

Spinning Winds and Tornadoes

Swirling wind effects are well documented, especially their role in disasters and trials by weather. One of the strangest spinning wind events took place in Washington, D.C., on August 25, 1814, as the British torched the city under siege. As the flames shot through public buildings, a tornado dropped into the middle of the city. "The darkness was as great as if the Sun had long set, and the last remains of twilight had come on," according to the British military historian George R. Gleig. The twister did major structural damage in the residential section, blowing off roofs, destroying chimneys, and flattening fences. But even more freakish, it maimed many of the troops in the attacking force. In fact, this striking tornado killed and injured more British soldiers than all the firepower the American troops used in the effort to protect the city.

Dust Devils and Lesser Winds: Spinning Winds of a Smaller Kind

On a small scale, sand, dust, and debris swirl across the Earth's surface in the form of *dust devils*. In the desert, they are called *whirlwinds, dust whirls,* or *sand whirls*. You can also see smaller versions of dust devils on windy days, as leaves, dust, and debris get caught up in vortices that spin across a flat parking lot, or even a swirling snowy dust devil when the wind whips up dry, light snow across a flat field.

Compared to their bigger brothers, hurricanes and tornadoes, dust devils do relatively little damage, occasionally flipping over a car or tearing off the roof of a house. In the desert, they appear in the middle of the day as winds blow around the local terrain. As the ground warms and the heat rises, strong updrafts create rising funnels of air. The reverse of a tornado (as you shall see), the spinning dust devil drags dust and sand into its rotating vortex, the wind speed increasing with height. Their diameters range from 10 feet (3 meters) to greater than 100 feet (30 meters), and they rise from less than 1,000 feet (300 meters) to several thousand feet high.

Combining some of the characteristics of a dust devil and a tornado, *gustnadoes* develop on the ground and throw dust several hundred feet upward. The very short-lived gustnadoes usually form near the leading edge of a thunderstorm, with winds strong enough to cause minor damage to structures. Many times these swirling patches are reported as fires, since the dust-filled winds look like smoke from a distance.

Mountain Turbulence: Mountain Dragons

Throughout the centuries, mountain winds have inspired wonder, respect, and awe. The high peaks dominate horizons, and their weather events seem downright magical. No wonder Greek mythology held two mountains sacred to Apollo and the Muses: Parnassus and Helicon, at the foot of which flowed two streams whose waters communicated the inspiration of prophecy and poetry. Of course, the Greeks also believed that their gods lived on Mount Olympus. Medieval stories told of knights doing battle with mountain dragons. And Viking legends told of giant trolls living in the mountains of Norway and Iceland. Even the Bible exhibits a fascination with the peaks; for example, God sent Moses to receive the Ten Commandments on Mount Sinai.

It is easy to see why certain peoples believed in the mountains' mystical and magical influence, in particular, the powerful mountain winds. If air at the bottom of a valley differs in temperature from the air above, a wind will blow. And if this wind gusts in the same direction as the prevailing mountain and valley winds, the winds will continually intensify as they travel through the thin corridors between mountain peaks. Add in storm winds, even a downdraft from a thunderstorm, and mountain winds seem supernatural.

As air flows across the mountains, the peaks act like rocks in a river, causing the winds to produce a series of ripples in the atmosphere. Forced up the peaks, the air tries to regain its balance by moving in somewhat stationary up-and-down waves, called *standing waves*. Then it settles back to a normal horizontal flow over the plains. But the standing waves that drop down the side of the mountains opposite the wind's direction (lee side) can create violent winds at the surface. Pilots flying aircraft through this wave call this phenomenon *clear-air turbulence.* In the summer, backcountry airplane pilots usually fly through the mountains during the morning and evening, when the air temperatures are relatively stable. Uneven heating of the rugged topography makes these winds too turbulent during the height of the day.

One remarkable kind of turbulence is called a *mountainado,* a spinning vortex that can cause major local wind damage in the mountains. Mountainadoes form as air flows over the uneven mountains and creates spinning winds extending to the surface; they form mostly in the summer or winter. The actual mechanism is similar to the formation of a tornado, but on a smaller scale. Warm air rises, while cooler air intensifies and strengthens the spinning motion of the winds aloft. Horizontal winds flow around an obstacle, such as the foothills, and run over the spinning vortex, actually pulling on the air as it runs by the central core of the vortex, causing the mountainado to intensify. Mountainadoes are most visible when they pick up and swirl light snow around the ground, with one reported at a height of 492 feet (150 meters). Winds inside a mountainado average 47 miles (75 kilometers) per hour, with some reaching up to 93 miles (150 kilometers) per hour. They have reportedly picked up roofs and sheds, and have even blown out windows.

• Mountain Airport Turbulence •

Hong Kong now has a new airport: the Chek Lap Kok Airport and its associated infrastructure. This airport has one of the world's best warning systems for aviation weather hazards, and for good reason: Lurking just off the airport's two runways are several small mountains that generate enormous turbulence problems.

The airport sits on the edge of Lantau Island, only a few miles from a half-mile-high mountain, with wind cascading within several miles of the approach and takeoff regions. Thunderstorms and high winds are common year-round, so turbulence and wind shear seem a certainty. The turbulence is the toughest issue. After all, it can occur in clear weather and can develop in less time than a microburst. Turbulent air flowing over a mountain shoots upward and intensifies, then plunges back down with erratically high winds that are difficult to predict. Not only that, but eddies can pop up as the air flows over an obstacle, like water flowing over rocks in a stream. To get a handle on the potential problems, officials are exploring computer-modeling solutions.

Ocean Waterspouts: More Watery Winds

The Literary and Scientific Class Book, written by Levi W. Leonard in 1827, tells what scientists knew about *waterspouts*, tall swirling patches of wind that gather water instead of dust and debris. "Waterspouts are often seen in calm weather; and the sea seems to boil and send up smoke under them, rising in a sort of hill toward the spout. . . . When these approach a ship, the sailors present and brandish their swords to disperse them, which seems to favor the conclusion that they are electrical. The analogy between waterspouts and electricity may be made visible by hanging a drop of water to a wire, communications with the prime conductors, and placing a vessel of water under it. In those circumstances, the drop assumes all the various appearances of a waterspout, in its rise, form, and mode of disappearing. It is inferred, therefore, that the immediate cause of this extraordinary phenomenon is the attraction of the lower part of the cloud for the surface of the water."

Awesome waterspouts have been reported, even in multiples, by sailing vessels across all the oceans and seas, on lakes, and even in rivers. *Nontornadic waterspouts* form from the interaction and mixing of rough and smooth seas churning up the water and winds near the surface. In this case, the water mixes and rotates with rising air warmed by temperature changes or wind currents—producing a swirling column of water. (Less pronounced nontornadic waterspouts that are associated with a few puffy daytime clouds are called *fair-weather waterspouts.*) Usually building up from the sea to the sky, both tornadic and nontornadic waterspouts can rapidly increase in height and diameter, and move slowly across the ocean surface for several minutes.

Spouts emerging from the clouds of thick, severe thunderstorms are called *tornadic waterspouts.* The first sign of a tornadic spout can be seen from the air but rarely from a boat. A dark spot on the ocean indicates that air is moving in a circle and upward. Then comes a spiral pattern of light and dark water, felt as a wind shift from a boat; at this point, the boater may also notice a funnel coming from an overhead or nearby cloud. Next, the swirling wind touches the ocean and is now easily visible from a boat. When the wind reaches about 40 miles (64 kilometers) per hour, it begins to kick up a spray of water, a circular pattern called the spray vortex. Next, the funnel extends all the way from the cloud to the ocean. You can usually see through the funnel, as it's really a thin cloud of tiny water droplets. Finally, the spray vortex weakens and the funnel gets shorter. This usually occurs when rain begins to fall from the parent cloud, the cool air from the rain cutting off the warm air feeding the waterspout.

• World-Class Waterspouts •

It is thought that waterspouts occur more frequently in the Florida Keys than anywhere else in the world. Around the Keys, especially from Marathon past Key West and on westward to the Dry Tortugas, there are usually 400 to 500 waterspouts a year, most commonly forming between 4 P.M. and 7 P.M. There are probably two major reasons for this proliferation of watery winds. First, the islands and shallow water produce a great deal of warm, humid air, and waterspouts need rising air currents in order to form. Second, the regular trade winds blow right down the islands, the winds causing the clouds to "line up," also encouraging waterspouts.

In fact, boaters in the Keys often underestimate the dangers of the waterspouts because the swirling winds are so common. But a waterspout does not respect people or things, and like a tornado or hurricane, deserves respect. Anyone who ventures out in a boat should stay well away from the spout and the swirling spray it kicks up after the funnel touches the water.

Supercells and Mesocyclones:
Where Tornadoes Incubate

Thunderstorms may form as single cells or multicell clusters or lines, but they often take the next powerful evolution into *supercells,* long-lasting thunderstorms that produce some of the strongest tornadoes, largest hailstones, and most dangerous winds. Many tornadoes originate from supercell thunderstorms, most forming in the Northern Hemisphere in the spring and summer months, when warmer air that the storm uses as fuel is more readily available. The warm, rising air in the thunderstorm helps to develop the storm by somehow imparting a slow rotation to the air. Such supercell storms have strong middle and lower spiraling, which create a *mesocyclone,* winds at the heart of a supercell that give the storms an extra punch. The mesocyclones need three conditions to form: warm, humid air at the surface, cooler air aloft, and something to push the warmer air upward.

Within the supercell, the warm, moist air rises like a hot-air balloon into the colder air, and the greater the temperature contrast, the greater the energy. Even more energy is added as the warmer and cooler air meet. In most thunderstorms, the resulting condensation of water vapor turns into ice crystals or water droplets; in the case of a supercell, the quickly rising air precludes the formation of ice crystals, even when the temperatures are below freezing. Because the small water droplets are packed like sardines in one place, they reflect a great deal of light, making these clouds look solid. As the rising air continues upward, it often overshoots the level where most clouds stop rising. This "overshooting" creates a dome-cloud at the top of the storm cloud. In most thunderstorms, a dome-cloud would look soft; in a supercell, because most of the water vapor has condensed as water droplets, the dome looks solid.

Meanwhile, the mesocyclone forms in the cloud, with the lower spinning air in the supercell tightening up into a long, narrow cylinder, then speeding up, similar to the way an ice skater tucks in his or her arms to spin faster. Most of the spinning energy comes from the right combination of wind speeds and the changing direction of the winds, both of which seem to twist the clouds into a mesocyclone.

And unlike normal thunderstorms, which dissipate when the supply of warm air is depleted, the mesocyclone's characteristics continue to fuel the supercell, allowing it to last for hours and move hundreds of miles. Though scientists do not understand the exact mechanism, the spin of the rising mesocyclone air seems to allow tornadoes to develop, adding to the twister's extra power. No rain falls from the bottom of a storm where the air is rising, which is why tornadoes form at the back of the storm, in an area away from the rain.

A good place to view a supercell is on a plain, such as the western plains of the United States (in the eastern United States, such storms are usually hidden by humid, hazy skies, or by other clouds). But the best place is from an airliner, though pilots stay far away from the storms: Jets fly at altitudes from 30,000 to 40,000 feet (9,144 to 12,192 meters); a supercell top can be 50,000 to 60,000 feet (15,240 to 18,288 meters) in height.

These towering cumulus clouds represent convective columns merging to form
a major thunderstorm with plenty of rain.

• For a look at information from the National Severe Storms Lab,

try linking to:

http://www.nssl.noaa.gov

• The Dryline •

There are seemingly hybrid, or specialized, ways in which thunderstorms develop, from sea breeze thunderstorms to the intriguing
formation of the dryline in the southwestern United States. Usually starting in New Mexico, a hot, dry wind blows down the eastern slopes of the Rocky Mountains, while moist air moves
northward from the Gulf of Mexico. The dryline forms as pockets
of hot air (thermals) rise and overturn in the lower parts of the
atmosphere, creating southwest winds to the west of the line; to
the east, the humid winds continue from the south.

The southwest winds are the strongest, and start to push

the dryline to the east. But the upper atmosphere moves faster, so the hot, dry air at about 1,000 feet (305 meters) overruns the warmer moist air below. This traps the moist air, creating what is called a capping inversion. If the cap breaks, moist air can explode upward at more than 100 miles (161 kilometers) per hour, growing into thunderstorms that are 50,000 feet (15,240 meters) high in minutes. And in some cases, these violent convection events can result in an outbreak of supercells, along with tornadoes.

Tornadoes: Twist and Shout

On May 31, 1998, wave after wave of thunderstorms hit upstate New York. Days before, the same line of weather had passed over South Dakota, spawning rapidly swirling, thick funnels of air called *tornadoes* and wiping out 90 percent of the small town of Spencer, South Dakota. By the time this line moved east and reached New York, a classic textbook weather scenario had developed. This strong cold front, with its collection of upper-air disturbances, arrived from the west just as a huge chunk of warm, humid air arrived from the south. To people on the ground, the thunder started first, continually rumbling as each storm wave rolled through. Each wave carried enough damaging hail, winds, and lightning to shut down the power for tens of thousands of people. The storms acted as natural pruners, breaking off tree limbs, beating down foliage, and tearing up sod.

But the rash of tornadoes—probably close to five—proved most memorable. In upstate New York's rolling hills and winding valleys, we think we're safe from twisters. But that day, we discovered that wasn't true. The morning light revealed more than a dozen destroyed houses. To add insult to injury, another tornado outbreak hit again two days later. Ultimately, the storms produced the worst outbreaks of tornadoes in our area in over half a century.

Tornado was first used in 1556 to describe violent thunderstorms in the tropical Atlantic Ocean. In the next century, the word referred to the swirling tubes produced by these storms. Since then, observers have seen the full range of their destructive power. There was the tornado-turned-waterspout that passed down the Ashley River in Charleston, South Carolina, on May 4, 1761, wiping out three ships of a British fleet assembling for a voyage to the mother country. Between April 3 and 4, 1974, some 148 tornadoes tore through an area from Alabama to Michigan. The longest tornado track on record was the Mattoon-Charleston tornado of May 26, 1917, whose swirling winds lasted for 7.5 hours and traveled 293 miles (471 kilometers).

Around the world, almost everyone has heard the term *tornado*. The swirling events pop up quite frequently in Australia, New Zealand, South Africa, Argentina, much of mid-Europe (from Italy north into England and Russia, with England experiencing the most), and occasionally in other places, such as Japan, eastern China, northern India, Pakistan, Bangladesh, Bermuda, and the Fiji Islands.

But none of these countries experiences the same high number of touchdowns as the United States. Our country has an average of 900 a year, with central Oklahoma having the distinction of experiencing more per acre than any place on

Earth. Tornadoes can form in any state, including Hawaii and Alaska, though they are rare west of the Continental Divide (there is not enough moisture to form the parent thunderstorms). "Tornado alley" in the United States is a broad swath of land extending from the Colorado Plateau down to the Gulf Coast. The area generates more tornadoes than anywhere else in the world, the combined result of the evaporation of moisture from the plains and driving winds of the jet streams.

Tornadoes are more likely seen in the Gulf Coast states in the spring. At this time the warm, humid tropical air from the south clashes with cold, dry polar air from the north, still lingering from the winter, to form major tornado-level storms along the boundary. Add solar heating on a warm spring day, and more tornadoes may result. As this boundary moves northward with the warmer weather of the summer, the tornado activity also moves northward, finally reaching into the upper Midwest by June and July.

These fascinating yet frightening and destructive weather phenomena hardly announce their entrance. The areas where tornadoes form are often indistinguishable from areas affected by average thunderstorms. Yet to one standing outdoors just before a tornado, the winds seem to pick up tremendously, howling through the trees and blowing debris in all directions. Sudden severe gusts of wind, downdrafts, and microbursts can also feel and sound similar, but their power stops at a certain point, whereas tornadoes simply grow into a menacing spectacle.

Tornadoes strike day or night, depending on their origin, their roaring vortex of wind as narrow as a rope or wider than a football stadium. Scientists can tell us which clouds have the potential to spawn tornadoes. But despite years of effort, scientists still cannot tell how far a tornado will travel, or even if or when a tornado will dip from the bottom of a storm cloud.

Most tornadoes originate in supercells. When a tornado is close to forming, the anvil-shaped top, part of the supercell storm cloud seems to develop a bulge on either side, "handles" indicating that the upward rush of air near the center of the storm is punching through to the tropopause. Another indication is the formation of mammatus clouds, the rounded, bumpy shapes on the bottom of the thundercloud. And finally, a "wall cloud" may hang from a larger rain-free cloud base, with heavy rain and large hail falling nearby.

Scientists do have a few answers to tornado questions. They know the resulting slender tubes form at the base of the thunderstorm clouds. The swirling clouds carry winds averaging less than 50 miles (80 kilometers) per hour, with parts of the tube potentially reaching speeds of up to 200 miles (322 kilometers) per hour. The majority of tornadoes have a forward speed of 20 to 40 miles (32 to 64 kilometers) per hour, but some have been known to travel more slowly and occasionally stall over one spot. They sometimes seem to travel in packs, the same clouds producing more than one twister at a time. The combination of wind force and pressure-drop suction easily demolishes buildings, uproots trees, and throws cars and trucks around like a giant with its toys. Although most tornadoes disappear in less than half an hour, some can last for hours and travel hundreds of miles before they disintegrate.

This wall cloud formed in Texas, just outside Austin. This particular thunderstorm did not produce a tornado, but it was a prime place for such twisting columns of wind to form.

• To find out more about the Tornado Project, try linking to:

http://www.tornadoproject.com/

and for more information on tornado alley, try linking to:

http://www.ou.edu/wx/CAPS.WWW

• Tornado Talk •

Tornadoes have been known to do some strange things, such as drive a piece of straw through a tree trunk; pick up people and animals, carry them hundreds of yards, and set them down safely; pick up a train, spin it around, and set it back on the tracks pointing in the opposite direction. And the list goes on.

If you're ever in an area known for its tornadoes, here are a few tips if you see an ominous cloud heading your way:

• If you hear a loud roar that sounds like "hundreds of freight trains" or "hundreds of bees" during a major storm, get to shelter immediately, especially a basement.

• Listen to your radio or television for tornado watches (when tornadoes seem likely), and head for the basement during tornado warnings (when the conditions are just right for a tornado).

• Look for a dark greenish or orangish cast to the sky. This *green sky* effect is thought to be caused by large amounts of ice suspended in the storm's upward-moving air (updrafts).

• Tornadic mesocyclones produce enormous amounts of lightning, so if you are listening to your radio, and the crackling from the lightning seems almost continuous, stay alert.

• If you see a dark lowering of a cloud base that seems to be rotating in a counterclockwise direction, usually in the wall clouds seen at the end of a severe thunderstorm (usually after a heavy rain and hail), stay alert. Also, if you see lumpy, cumulonimbus mammatus clouds near the end of a severe thunderstorm, stay alert. These usually indicate the extremely unstable conditions that can precede a tornado.

• And if you're a storm chaser, beware of the *bear's cage*, the part of the storm with the most intense rainfall, lightning, and hail, usually north and east of the mesocyclone region containing the tornado (chasers usually approach a storm from the southeast through west quadrants). If you do travel through the cage, you may come face-to-face with a tornado, with little or no warning.

• Tornado Classes •

Tornadoes are classified by the destruction they leave behind. The first person to classify tornadoes was the University of Chicago's Dr. Theodore Fujita, one of the world's leading experts on tornadoes, who noted the pattern of damage compared to the wind speeds. Though every tornado should be considered major, since they all have potential for extensive damage and loss of life, here are a few statistics: F0, F1, and F2 tornadoes are the most common, and fewer than 2 to 3 percent of all tornadoes are F5 tornadoes. All the following numbers are approximations.

Tornado F Rating (F scale)	Winds (miles/kilometers) per hour)	Length (miles/kilometers)	Damage
0	40–72/64–116	under 1/1.6	light
1	73–112/117–180	1–3/1.6–4.8	moderate
2	113–157/181–253	3–10/4.8–16	considerable
3	158–206/254–331	10–31/16–50	severe
4	207–260/332–418	31–99/50–159	devastating
5	261–318/419–512	99–315/159–507	incredible
6*	319–380/513–611	315–999/507–1,607	inconceivable

* An F6 tornado is only theoretical; no tornado has ever reached such levels, and it is hoped none ever will.

• Smokestack Tornadoes •

If you're lucky enough to spot one, you can catch a plume of heat flowing from the tops of industrial smokestacks. These are also perfect places to see a small version of a tornado in action: smokestack tornadoes. The rising thermal plume often interacts with light or moderate winds, producing a swirl of smoke from the stack. These twirling tornadoes last for only a few seconds to a few minutes, as the thermal plume is not large enough to sustain the spinning.

8

Mountains and the Weather Above the Clouds

We decided someone was watching over us: An early fall blizzard in 1997 dumped up to 3 feet (1 meter) of snow on the east side of the Rocky Mountains, paralyzing the city of Denver and closing highways both north and south along the Front Range. We had just passed through the area two days before, and for us now in northern Arizona, the weather was sunny and warm. We had missed the storm: Stranded motorists had to be rescued from their cars by helicopters and snowmobiles. Airports were closed and commerce ground to a halt. Yet the excitement was over in a short time, as the snow along the range was mostly gone within a few days, evaporated by a warm, dry wind flowing down from the mountains. From one extreme to another.

Mountains, Not Molehills: Getting in Weather's Way

A person's perception of "mountains" is usually based on those they live closest to, but the peaks vary greatly. There are the younger, taller, and more angular mountains, such as the Himalayas, Alps, and Andes, all born from the recent (geologically speaking) violent collision between continental plates. The millions-of-years-old uplifts continue today, as chunks of crustal plates plow into one another, raising the land in each mountain range inches per year. The Rocky Mountains, rising about 80 million years ago—and continually uplifting for 40 million years during a time called the Laramide orogeny—still exhibit some of the angularity of their youth. There are the well-worn mountains, such as the Appalachians, old and rounded for more than 400 million years by eroding agents such as wind, rain, and ice. And there are the shield-type volcanoes such as the Hawaiian Islands, or the peaked volcanoes such as Mount Hood, each with its own characteristics, depending on the lava and ash thrown from its belly.

And literally caught in the various mountains are specific weather patterns and phenomena, the likes of which are seldom seen elsewhere. The reasons are simple: Mountains are great blockers, pushers, and trappers of air, massive chunks of uplifted land getting in the way of weather systems.

• Drop the Wind •

As can be expected, mountains have a tendency to change the flow and actions of moving air. The Rocky Mountains, the longest north–south mountain barrier in the world, also stand in the way of the prevailing west-to-east winds; this creates large-scale weather patterns, especially strong winds as they break through the mountain range. These winds sweep down the eastern side of the Colorado Rocky Mountains and into Boulder, a city that sits where the mountains meet the plains. The resulting winds can gust at 100 miles (160 kilometers) per hour or more, ripping roofs from buildings, snapping power lines, shattering windows, and sandblasting paint from cars. The damage from the winds averages about $1 million per year, and one storm that occurred on January 17, 1982, cost more than $10 million. It is interesting to note that Fort Collins and the southwest Denver suburbs are rapidly expanding into the area of the winds. And the winds won't go away, no matter how many people move there.

Clouds from an advancing cold front get caught on the Rocky Mountains along the Continental Divide. From the vantage point of an airplane, it's easy to see how the clouds dump so much snow on the windward side, and dry air falls down the lee side of the mountains.

Mountain Winds: In the Shadows

The *rain-shadow effect* has nothing to do with shadow and light. It occurs as moist air runs up the windward side of a mountain, where it is forced upward and spread out, forming and generating clouds that eventually drop rain, fog, or snow along the mountainside. It is almost as if, by going up the mountain, the moisture is squeezed from the clouds. By the time the clouds reach the top, most of the precipitation has fallen, which is why most of the snows fall on mountaintops. As the now dry air sinks at the peak, it tumbles down the other (lee) side of the mountain slope, naturally warming, heated by the compression of higher air pressures at lower altitudes.

This rain-shadow effect is the reason for huge rainfall or snowfall amounts on the windward side of the mountains, and the dry deserts just beyond the lee side of the mountain ranges. For example, in the Sierra Nevada of the western United States, more than 100 inches (250 centimeters) of rain and snow fall on the windward side of the mountain range; to the east lie the dry Death Valley and a great desert zone covering an area of eastern California and all of Nevada.

What do the "shadows" have to do with mountain winds? The rain-shadow effect produces the strongest winds on the lee side of the mountains. For example, north of the snow-filled Himalayas, the Gobi Desert sits, the result of dry winds falling from the lee side of these tall mountains. Some of the wettest places in the world are on the sides of coastal mountains facing the prevailing winds. One is Hawaii's Mount Waialeale, a mountain with an average of 360 rainy days per year. The winds over the ocean pick up moisture and drop it along the slopes of the volcano; on the other side of the peak are dry grasslands resembling the western United States, complete with cactus.

• Mountain Wind Cycle •

Did you ever notice how the winds affect your campfire in the mountains? The fire blows one way all day and the opposite direction at night. This daily cycle of wind is similar to the land and sea breezes felt along rivers and oceans. In the case of the mountains, the Sun warms the slopes during the morning, heating the air near the ground. Because it is lighter than the surrounding cooler air, the warm air rises up the mountain slope. The cooler air from above sinks to fill the space. Then after heating, the air flows upward in a continuous cycle; the resulting winds are called *valley winds*.

At night, the opposite happens: The air above the mountain slope cools, sinks, and travels back down into the valley, creating winds called *mountain winds*. Similar to the rain-shadow effect, the valley winds often drop moisture as they ascend the mountain slope. Thus, the mountain winds at night are often drier as they flow back down into the valley. Amazingly, these winds are vital to flora and fauna of many of the higher mountain ranges, as the winds blow snow off some slopes and build up drifts on others, determining where plants will grow.

• For more on one of the windiest mountains in the world,
Mount Washington in New Hampshire, try linking to the
Mount Washington Observatory:
http://www.mountwashington.org/

Even fair-weather clouds can get caught on a mountain, a typical sight in the
Blue Ridge Mountains of Virginia.

Famous Winds: Among the Snow Eaters

Depending on the season (most often in the winter), the wringing out of warm, moist air from the rain-shadow effect creates some of the most famous warm winds—also called "fall" winds—on the lee side of the mountains. As the wind drops down the lee side of the mountain, it displaces the cooler air, causing a dramatic rise in temperature in a few minutes, as much as 40 degrees in 15 minutes. For example, the *foehn*, or *föhn* (the original name for these winds) blow in Alpine valleys, such as those on the east side of Switzerland's Alps. Similar winds are called the *zonda* in Argentina, *halny wiatr* in Poland, and *koembang* in Java. In the United States, the *Santa Ana winds* descend from the inland high deserts, skirt through the nearby mountain passes, and then spread across the southern California coastal plain. And the *chinook* (literally, "snow eater") winds tumble down the east side of the Rocky Mountains.

The list of other local rain-shadow-related winds around the world is huge; thus, only a few will be mentioned here. Many of the winds have their own local nonscientific name, and not all of them have exactly the same effect, but most are close. For example, one of the more popular is the *mauka* in Hawaii. This is the name not of a new hula but a cool breeze that occurs during the night, as winds descend from the nearby volcanic mountain slopes.

Another type is the dry, often gusty *bora* winds that fall down the lee side, causing the temperatures to drop, not rise. They are named after the winds that blow in the winter over the Dalmatian Mountains in the former Yugoslavia and toward the Adriatic Sea, events normally lasting twelve to twenty hours, or for six to seven days at least once each winter. Bora-type winds are found all around the world: winds on the Black Sea coast of the Crimea, in the Apennines of Italy, and the fjords of northern Norway. Analogous winds with local names occur in the Kanto Plain (inland of Tokyo, Japan) called the *oroshi*; and winds at Crossfell in the northern Pennines of England.

The *mistral* winds are cold, dry northerly, or northwesterly, winds flowing into the Mediterranean Sea and skirting along the northern coast from Ebro to Genoa. These winds toss the sea, creating huge waves—mistrals reach speeds in excess of 86 miles (138 kilometers) per hour—and are most intense usually around the coasts of Languedoc and Provence, especially around the Rhône Delta. They are caused by the sinking of cold air forming over mountains, gaining strength as they are funneled through gaps in the peaks of the Pyrenees and Alps.

Even the colder regions of the world are not immune to such

Cumulus mediocris clouds are found worldwide, except over the continent of Antarctica (where the cold surface does not allow convection to occur). These billowy clouds over the Cannonsville Reservoir in New York formed in the early afternoon, as the ground generated enough heat for convective activity to begin. They are too small to generate any rain.

winds. On the ice sheets of Antarctica and Greenland, strong *drainage winds* move down the slope of the ice surface. These drainage winds (cold air that flows under the influence of gravity from higher to lower regions) are then funneled through the coastal valleys. These strong, swift-moving cold winds stir up the snows, creating powerful blizzards lasting for days at a time.

• The Attraction of Mountains •

Mountains can generate not only snow or rain records, but certain unique weather phenomena. Take, for example, the mountains around Boulder, Colorado: According to the National Center for Atmospheric Research, the area has the second highest number of annual lightning strikes in the United States (the first is in Florida). The frequent lightning flashes are brought on not only by the storms churned up over the mountains but also by the mountains themselves, it is thought that the iron in the mountains' rocks attracts the lightning.

Upslope Effects: What Goes Upslope

Even a small mountain chain can create a sort of Bermuda Triangle for weather forecasters, a zone in which weather is almost predictable but is then sabotaged by mysterious forces. According to Mark Hanok, an Otego County–based meteorologist who publishes the *Western Catskills Weather Gazette*, and a forecaster for radio station WRKL, upstate New York is one of those places. During most winters, lake-effect snows sweep down southeast from the Great Lakes region. As cold northwest winds swing down over the lakes, they pick up moisture, usually dropping it along a region that stops just short of the cities of Corning, Binghamton, and Cooperstown. But sometimes, those who live beyond this imaginary sinuous demarcation are not that lucky. Snows still drop beyond this point because of the *upsloping effect*, the result of the air's cooling and condensing during its rise over the Catskill and Pocono Mountains. Here, the smaller mountains create the larger weather events.

South-central New York is not the only region with a relatively small mountain chain that carries a big weather punch. Seattle's great rainfalls are often due to the *Puget Sound convergence zone*. In this case, the eastward-flowing air from the Pacific meets the Olympic Mountains. The winds take the easiest paths, flowing to the north and south sides of the mountains, with both airstreams converging just beyond them. The air also rises and cools, increasing the precipitation.

Other upslope effects occur in the unlikeliest places. Along the Front Range of the Rocky Mountains, where the mountains meet the plains, the term *upslope winds* also means bad weather. Humid air, blowing westward from New Orleans to Denver (which is strange, as most air flows from west to east) rises along the gentle slope of the plains. It cools at the rate of about 5.5 degrees for every 1,000 feet (305 meters) it ascends. Clouds and rains develop as the air rises. Depending on the temperature of the ground the air runs over, upsloping can cause strong weather activity: fog on the plains, rain and snow in the hills and mountains of the western plains, and snow in the eastern Rocky Mountains.

• Humans and Mountain Air •

Though steep slopes, intense cold, and air with low oxygen all distinguish higher mountains, people still brave the barren areas, either living nearby or taking excursions into these regions. Life at higher elevations is not easy. The bodies of the Andean Indians of South America exhibit the telltale characteristics of adaptation to the high mountains—more oxygen-carrying red blood cells and larger lungs compared to those of lowland peoples.

Probably the most famous mountain associated with people is Mount Everest, the tallest mountain on Earth at 29,022 feet (8,846 meters), and the peak many people want to climb to "reach the top of the world." Each year the mountain claims many lives. For example, in 1996, the most lethal year so far, one person died for every six and a half who reached the summit, most dying as the

result of falls and avalanches. Studies recently have been conduct-
ed to determine if oxygen deprivation (part of high-altitude sick-
ness) is contributing to the accidents, the lack of oxygen causing
the brain to react slowly to life-threatening events.

Mountain Fog: The Shadow of Your (Mountain) Smile

Fog in the mountains allows light and shadow to play tricks on
the observers. In the Hartz Mountains of Germany, one can
meet the *Spectre of the Brocken* (or Brockenspectre), named
after the area where the event is commonly reported. If you're
high on the mountain with your back to the Sun and a cloud of
fog is just about to surround you, your shadow can penetrate the
mist for some tens of feet. Depending on where you are standing, your arms and legs seem stretched into long shapes, as if
you were an enormous Gumby doll. Often, the concentric rings
of a glory or a fogbow surround the upper parts of your shadow.

A human shape is not the only one a shadow can assume in
the mountains. The mountains themselves can form triangular shadows in the sky—even if the mountains are not triangular in shape—especially if viewed at sunset and where there is
a significant amount of dust in the sky. The National Center
for Atmospheric Research, located high above Boulder, Colorado (near the Flat Iron Mountains), is a prime place to spot
the shadows. One sees them especially on hot summer days,
when the winds from the mountains race downslope, churning up the dry dust to the east. Here, the airborne particles
make a fine backdrop for the mountains' triangular shadows.
The reason for this is perspective. Viewed from the top of a

mountain, the shadow seems to converge in the distance. Consider your own shadow at sunset. It's elongated, so stretched out that it essentially comes to a point at the top of your shadow head. This is similar to the peaked mountain's distant shadow.

Airlight: Airlight, Air Bright

It is no wonder the song "America, the Beautiful" contains the words, "purple mountains' majesty," as certain light does create a purplish hue around peaks. In reality, most of us see blue in the mountains. The blue tint of distant mountains is common, especially in peaks that extend for tens of miles, such as the Blue Ridge Mountains of Virginia and North Carolina. Look carefully as you travel past them. On a clear day, you'll notice that the farther you look, the bluer the mountains. This bright blue veil is called *airlight,* caused by air molecules and particles scattering sunlight between the viewer and the mountains.

You can see airlight only during the day, most often when the Sun is overhead. In a very broad sense, you might even consider sunlight itself to be a kind of airlight, but it remains somewhat unspectacular because there is nothing to contrast it with in the blue sky.

As you look farther at the successive mountain peaks, you are looking at greater distances through the air. At such distances, the airlight (and the contrast) increases, lending the farthest mountain the lightest appearance. The greater the distance you look, the more the light from the mountain is scattered until the atmosphere becomes opaque and the mountain range "disappears" from sight. In fact, this is why you can see only a limited distance on Earth, no matter how clear and crisp the air is on a certain day. Airlight can be so bright, especially when you are looking toward the Sun, that distant mountains "disappear" during the day, only to show up again when the Sun sets.

Airlight is not the only mountain "glow" phenomenon. One can also encounter the scattering of light called *alpenglow*. You can see these pink, orange, or gold tints on the highest peaks for a few minutes after sunset. They often show up in photographs of high peaks around the world. As the Sun nears the horizon, the greater number of dust and aerosol particles near the Earth's surface scatter all colors except those at the red end of the spectrum. At first, the particles scatter the direct light of the setting Sun, causing the stark, reddish colors. As the Sun sets, the glow continues for a few more minutes, as the light reflects off any clouds near the setting Sun.

Unlike the blue airlight, *haze* is brown, white, or gray, and is smaller than cloud droplets. When cars, industry, or commercial enterprises emit smoke, dust, and other particles, mostly of pollutants, haze results. Denver is a prime example, as the city is virtually in a bowl next to a mountain range, a perfect setup to catch the pollutants. Likewise, Los Angeles is essentially a flatland with an inland mountain range that can hold the winds from the ocean in the basin, thus preventing the area from getting rid of its particulate haze. The reason for reduced visibility in haze is the particles' ability to scatter visible light.

Airlight is easy to see in the ridged, rolling hills of Virginia's Blue Ridge Mountains. Also in this image are the faint beams of crepuscular rays.

Volcano Weather: From the Belly of the Earth

Volcanic gases formed much of the Earth's early atmosphere. Rocks older than 3.5 billion years old show there was much more carbon dioxide in the early atmosphere than today. Later on, plants began the continuing process of photosynthesis, releasing oxygen and changing the overall composition of the atmosphere. In other words, we should thank the *volcanoes*. If they hadn't initially exploded, there would probably be no atmosphere, and no life as we know it on Earth.

These giant volcanic blemishes on our planet's surface still release gases into the atmosphere, though not as extensively as they did during the beginning epochs of the Earth's history. Modern volcanoes are no doubt similar to early ones, spewing carbon dioxide, water vapor, and lesser amounts of other gases such as nitrogen, methane, and argon, into the atmosphere.

Even though they've been relatively quiet of late, a major eruption could change our climate.

Consider the 1883 eruption of Krakatau. This volcano burst forth with ash and dust in the loudest explosion in recorded history, a boom heard over 2,000 miles (3,218 kilometers) away. Other volcanic events caused climate and weather changes, including the persistent dry "fog" over Europe for much of 1783 and 1784, thought by Benjamin Franklin to have been caused by the huge eruption of the Laki volcano in Iceland. There were also major volcanic eruptions around the world the year before 1816, called the "year without a summer."

More recently, Mount Pinatubo in the Philippines erupted on June 15, 1991, ejecting some 15 to 20 million tons of sulfur dioxide gas and ash into the air. This influx of material took three weeks to spread throughout the atmosphere, from west to east across the globe. In its wake, space shuttle astronauts could see a hazy blanket over the Earth, the likes of which they had never spotted before. The blast of particles into the atmosphere also blocked the Sun's incoming radiation, lowering the mean global temperatures by one-half- to one degree Celsius for the next several years. In a perverse way, explosions like those of Mount Pinatubo can reverse the trend of global warming. Of course, there are better ways to reach this end.

• Almost the End of the Human Race? •

Massive volcanic explosions can have extreme consequences. One such eruption occurred close to 74,000 years ago, toward the latter part of the Ice Ages, during a time called the Wisconsin glacial advance. This huge volcanic burst occurred on the Indonesian island of Sumbawa, from a large volcanic mountain called Mount

Toba. This eruption is thought to be one of the most violent in the past million years and occurred just about 3,000 years after humans (*Homo sapiens*) made their way out of Africa. The explosion added climate problems to the already glacier-covered cold Earth. The dust, ash, and particulates blanketed the global sky, creating a "volcanic winter" that eliminated some areas of vegetation and changed others. Some scientists believe this colder weather and change in food sources may have killed off various precursors to the human race.

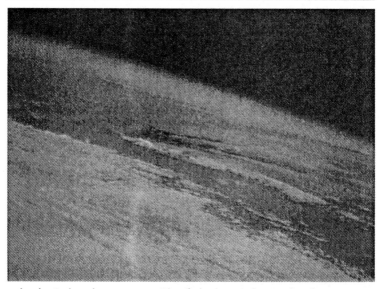

A volcanic dust plume over an unidentified volcano in Russia, taken during space shuttle mission STS-68 in 1994.

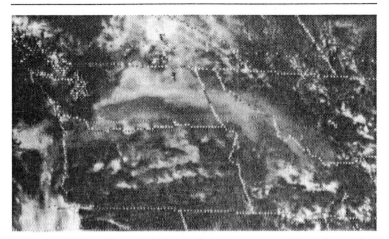

A volcano such as Mount St. Helens in Oregon can put out millions of tons of ash, gases, and dust into the atmosphere, as it did in 1980. If the conditions are just right, these particulates and gases create some striking weather phenomena, including brilliant sunsets and perhaps even a change in our global climate. (photo courtesy of the National Oceanic and Atmospheric Administration)

Volcanic Effects: A Palette of Colors

Erupting volcanoes can generate weatherlike activity in the local atmosphere. The birth of the island Surtsey, just south of Iceland in the North Atlantic Ocean, was a classic case. The volcano spewed molten rock and hot ash into the air, and giant clouds of steam and debris. As observers watched, the smoke clouds metamorphosed into a regular thunderstorm. As with typical atmospheric cloud lightning, charges built up within the volcanic cloud and reacted to charges on the ground, creating "volcano lightning." In addition, the clash of steam, ash, and dust led to mini-tornadoes around the volcano.

Volcanic eruptions can also create some of the most brilliant skies imaginable at sunset and sunrise. If you look toward the horizon at the setting Sun, you are looking through many layers of the atmosphere. During "volcanic" years, you are looking at the results of the sunlight refracting off the excessive amounts of volcanic particles. For example, the 1991 eruption of Mount Pinatubo in the Philippines produced lengthy, spectacular red-colored sunsets for several years afterward.

• Once in a Blue Moon •

In September 1950 the United Kingdom and the nearby continent were treated to a blue-green Moon. Strong westerly winds in the upper troposphere and huge forest fires in the United States' Rocky Mountains and Alberta, Canada, pushed dust and ash all the way to western Europe, the particulates scattering the light and tinting the colored Moon. Such events may be the real meaning of "once in a blue Moon," though many astronomers (erroneously) use the phrase to denote the second of two full Moons in the same month.

• Earthquake Weather? •

The volcano is a partner-in-crime with the earthquake, since both events commonly occur along the shaky boundaries of the planet's crustal plates. Earthquakes can also produce strange electromagnetic disturbances several hours or days before the quake itself.

The January 1995 Kobe earthquake in Japan killed around 2,500 people and caused millions of dollars in damage. During the quake, many Kobe residents saw shimmering lights or flashes in the sky similar to lightning. Others saw strange bluish-orange lights.

Other earthquake events have generated reports of sparkles, streamers, and ball and regular lightning. The colors range across the spectrum, mostly white, orange, and blue. Scientists believe such effects result from the squeezing of rocks, which turns mechanical energy into an electric charge. The intense electric fields then ionize the air, producing the lightninglike glows.

9

Dry Skies of the Desert

Huge thunderstorms seem more harrowing on the desert, probably because the wide horizon allows you to take in the entire monster storm. A few years ago, we were heading west along Interstate 40 toward Flagstaff, Arizona, and heard the crackle over the radio, a good indication that it wouldn't be too long before we ran into a storm. Coming over a rise just past Petrified Forest National Park, we almost stopped the car and turned back. There, not a mile from us, and stretching from north to south, was the toe of a cold front. As the tip of the front churned up the dusty soil of the desert, it looked as if we were about to hit a wall of sand. As we continued into the toe, the winds in the forefront rocked the car. Our surroundings took on an orangish tint, as the light reflected the desert's yellowish orange sand particles around the car. Unexpectedly, the winds slowed and the sky cleared of sand. And there it was: a gigantic thunderstorm supercell heading our way—the gods of thunder riding across the flat, dry desert, churning up dust and debris, with nothing to cloak them, nothing to stand in their way.

Dusty Deserts: Why It's Dry

Long ago, a science fiction writer revealed a secret: Always put your hero on a cold planet. After all, he or she could put on more layers of clothing to survive the frigid weather. But in a hot and dry climate, you can't always cool down. Your hero can't necessarily find water to replenish what has evaporated. (Thus, the cold of space became colder as more stories were written with this in mind.) But in reality, humans are more versatile. In fact, some evolutionary scholars think that our very species first evolved around the hot, dry savannas of Africa, adapting to the oppressively hot temperatures of the region.

Just what is a *desert*? For most of us who watched such movies as *Beau Geste* and *Lawrence of Arabia* when we were young, it means dust storms, oases, and mirages. If these were the true distinguishing features of a desert, however, almost everywhere in the world would count as a desert at one time or another. Summer heat and winds pick up dust and debris, swirling it in parking lots like a dust storm. No matter what the rain situation is, you can always find pockets of lush vegetation in most geographic areas. And the sweltering heat of a mid-summer's day can change a normal macadam road into a mirage.

Deserts, however, are officially defined as areas of low precipitation and high evaporation rates, a region with less than 10 inches (25 centimeters) of rain per year, and in some years, no rain at all. With about 90 percent of the sun's energy reaching the ground, daytime summer air temperatures are often greater than 100°F (38°C), stretching to 120°F (49°C) in the warmer spots. As the Sun sets, the temperature plummets, and because of the lack of moisture, there are no clouds to hold in the warmth. Temperatures can drop by more than 60 to 80 degrees during the night. The winter temperatures can drop even more, with places like the Gobi Desert recording as low as −5°F (−21°C).

And some unlikely places actually count as deserts. While the sandy lands of the Sahara and the Gobi are clearly deserts, cold deserts abound at high mountains, where precipitation is also low and the land mostly barren of vegetation. The colder regions of the Arctic and Antarctica are classic deserts. Even the oceans, such as the waters surrounding the Hawaiian Islands, have been called a desert. For there, too, precipitation is low and evaporation rates are high.

• *For more information on the deserts, try linking to:*
http://www.webstories.co.nc/focus/deserts/

• Driest of the Dry •

The Atacama Desert of northern Chile, South America, is the driest place on Earth. And although the average rainfall is 0.003 inch (0.08 millimeter), years can pass before a rainstorm rolls through the region. The driest place in the United States is Death Valley, California, which gets a yearly rainfall of 1.63 inches (4.14 centimeters).

Sandstorms and Dust Storms: Natural Abrasion

On a grand scale, there are the *sandstorms*, hundred-feet-high walls of sand thrown into the air by high winds. Since no vegetation or water holds the sand particles in place, sandstorms happen quite frequently. Some, such as those that form in the Sahara Desert, carry tons of sand into the air. The sand particles carried by the strong winds of a sandstorm rarely reach more than 50 feet (15 meters) above the ground. Yet the storms are strong enough to carve the desert into hills of sand called dunes, and shift the sand mounds around the desert. In particularly violent storms, the winds can move a dune 65 feet (20 meters) in a single day, burying rocks, fields, and eventually even towns.

A sibling to the sandstorm is the *dust storm* (also called a duster or black blizzard), in which strong winds stir up smaller particles of silt and clay. They can occur in deserts but also form in more temperate latitudes after a major drought. Dust-storm clouds may be up to 10,000 feet (3,050 meters) high, the soil and dust mixing together to form a virtual moving wall that can trek over thousands of miles; the winds are usually greater than or equal to 30 miles (48 kilometers) per hour. A dust storm is present if visibility is reduced to about half a mile (1 kilometer); if a dust storm is severe, the visibility falls to one quarter mile (0.5 kilometers). In the most severe dust storms, visibility is only a few yards and the area is reduced to darkness.

Typical dust storms form along the Great Plains in the United States behind a cold front or dryline. As the front moves

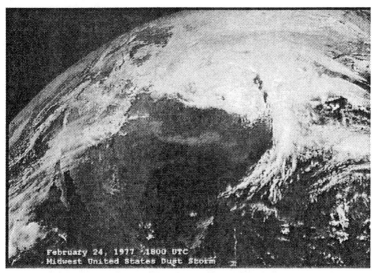

February 24, 1977 1800 UTC
Midwest United States Dust Storm

Massive dust storms do not always have to occur in the deserts. This image shows a major dust storm racing across the midwestern United States, a combination of drought conditions and high winds. (photo courtesy of the National Oceanic and Atmospheric Administration)

through, the strong winds at its nose lifts the loose topsoil off the ground, pushing up a wall of dust high into the air. As the front plods along, it carries the wall with it, adding more material from the surface. In the deserts, the winds causing these storms are called the *haboob*, strong squall winds that kick up mountains of sand.

• Seasonal Desert Winds •

On a grand scale are the seasonal winds, once understandably thought to be gods or other supernatural beings. Now meteorologists know that most of these winds are seasonal, brought on by various combinations of hot and cold. And today, these familiar wind patterns abound in the desert areas in the world.

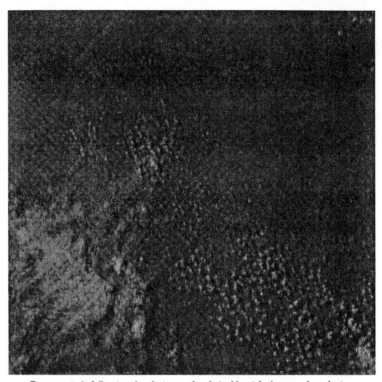

Desert winds following the drainage clouds in Algeria's desert, taken during space shuttle mission STS-73. (courtesy of NASA)

The *sirocco* is typically a dry, hot, irritating wind that blows out of the Sahara and Arabian deserts. Each February through April, it flows hundreds of miles across northern Africa to the Mediterranean Sea. As the winds move across the Mediterranean, they gather moisture. By the time they reach Europe, the winds are warm and humid.

The *khamsin,* or the winds from the Sahara Desert of Africa, bring hot, dry air to southern Egypt, often devastating crops. The *gharbi* winds also start in the Sahara Desert, gathering up moisture and depositing heavy rains over the northern and eastern Mediterranean. The resulting precipitation mix of red sand, dust, and water is called the "red rain." Away from the Sahara Desert, the *berg* winds flow from the interior of southern Africa toward the coast, bringing days of scorching temperatures as the hot, dust-filled winds fill the air. The *brick fielder* winds are also hot, dry, dust-filled winds blowing from the deserts of Australia toward the city of Sydney, and then to the rest of the southern coast.

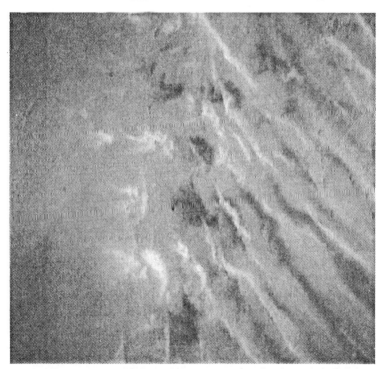

Wispy clouds form in the Sahara Desert, contrary to the often popular belief that clouds don't form in the desert; taken during space shuttle mission STS-56. (courtesy of NASA)

Global Effects of Sandstorms and Dust Storms: Around the World in (Less Than) Eighty Days

In early spring, over the deserts of central Asia and China—especially the Gobi—winds pick up huge clouds of dust. The winds blow the sands east, often reaching Hawaii and the other Pacific islands within the strong winds' paths. Don't conclude that the dust covers the islands like volcanic ash. It doesn't, though it does produce colorful sunsets.

The great Gobi is not the only desert that spreads its sands across the world. The sands of Africa's deserts also travel to other continents, sometimes reaching as far north as Europe and as far west as the Americas. For instance, the sandstorms and rising hot air in the Sahara Desert can carry the dust up to 15,000 feet (4,572 meters). In the fall of 1997, westward winds carried the sands across the Atlantic, actually causing hazy skies in the northeastern Caribbean. At the same time, scientists found Sahara dust in parts of the southeastern and eastern United States, and hypothesized a connection between the sands and certain hazy days in the Grand Canyon of Arizona. Similarly, dust storms originating over southeastern Australia carry the dusty "red snows" across the Tasman Sea to New Zealand's Southern Alps. (Dust in the central United States can also be carried by the prevailing winds to the Atlantic coast, especially during dry times in the plains.)

Not that the influx of particle-filled winds from the African deserts is bad. In fact, they often benefit the areas they reach. For instance, the dust fertilizes the Atlantic Ocean, providing iron-rich soil for ocean plankton and krill. The main "dust routes" are rich breeding grounds for fish and other marine life. In the eastern Amazon rain forest, more than 13 million tons of dust fertilize the region annually. One scientist suggests that these dust particles are chemically alkaline (basic), which may dilute sulfuric acid, a main component in acid rain. Such dust may reduce the effects of acid rains so prevalent in the eastern United States.

Desert Storms: Where Does It All Go?

Rainstorms in the desert don't happen often, but when they do, the resulting influx of water can be devastating. Usually a flash flood causes the most damage, since a thunderstorm drops heavy rain onto a desert unable to absorb such great torrents of water. Desert flash floods have eroded banks, caused landslides, and collapsed houses as they race through the flat, sandy lands.

One curious desert storm phenomenon is called virga, a wispy sheet of rain that falls in the distance but never seems to reach the ground (virga can also be found in more temperate regions on hot, dry days). As the raindrops fall through the air, they encounter a layer of dry, warm air before they hit the ground, and the rain evaporates. Scientists think something similar happens on the hot, dry planet Venus, except there the raindrops are made of sulfuric acid.

• Pushy Desert Winds •

Wind-propelled rocks are some of the strangest sights in the desert. Called *sliding stones*, these rocks make a path across arid terrain almost as if in search of other territory. They also leave strange tracks on playas, flatlands in deserts usually covered with salts and sediment. There are a few theories as to why these rocks move. The most popular theory is that the rocks move in response to wet, stormy weather. As rain covers the parched ground, winds blow the rocks, causing them to scamper across the desert floor. In the winter, as thin sheets of ice form on certain playas and deserts, winds can also drive the stones across the slippery ground.

Droughts: When the Land Turns to Desert

Patricia's mother once described the years of the 1930s Depression era as "one disaster after another," including the northeast flood of 1936, which collapsed bridges and carried whole houses down the Susquehanna River near her home. She was quite young when one of the more famous natural disasters hit: "the" *drought*. It started in the southwestern United States in 1930, slowly spreading east and northeast, becoming progressively worse every year until it peaked in 1936. High temperatures and hot winds evaporated what moisture was in the soil and sapped the nation's water resources. The Midwest was nicknamed the "Dust Bowl," the land becoming worthless to farmers, the dry soil torn apart by the winds and blown across the dry land. In 1934, wheat, corn, and other grain crops neared total failure. And what the heat didn't take, the insects did. Everyone watched as the price of vegetables, flour, and other foodstuffs rose steeply, not only in the hardest-hit regions but around the entire nation. Just to get enough to eat, people planted backyard gardens everywhere around their neighborhoods. The heat rose to record levels in many spots in the Midwest and East, and it is estimated that more than 15,000 people died in drought-related tragedies. Not until 1940 did relief arrive, when the rains finally fell.

Periodically, humid areas, especially mid-latitude regions in which a certain amount of rain is expected to fall, can go dry, experiencing droughts. The definition of a drought varies from country to country. In the United States, the term *drought* includes an area receiving 30 percent or less of its usual rainfall over about 21 days. In Australia, it is an area receiving less than 10 percent of the normal annual rainfall, and in India, an area receiving less than 75 percent of the normal annual rainfall. Regardless of the definition, droughts are powerful. The dryness of a drought kills crops, and bakes and hardens the ground, optimizing flash-flood conditions by not allowing water to seep into the soil. Water shortages decimate wildlife and livestock. Winds whip up the dry soil and often create wildfire conditions. The heat of more severe droughts can kill both animals and humans.

Droughts have complex origins. Recent computer prediction models link droughts with ocean surface temperatures, including the patterns that develop across the Pacific Ocean and off the west coast of South America—the most well known being El Niño. Other meteorologists suggest a loose association with the sunspot cycle, in which our Sun produces more sunspots and solar storms every eleven years. For example, in

1988, during the peak of a sunspot cycle, meteorologists noticed a fluctuation of the ocean temperatures and a shift in the jet stream. This, in turn, seemed to stall a high-pressure system over the middle of the United States. This entrenched system stopped any rain-carrying low pressure from moving to the east, creating the worst drought since the 1930s.

• For more information on droughts, try linking
to the National Drought Mitigation Center:
http://enso.unl.edu/ndmc/

• Drought Records Everywhere •

The world has had its share of parched weather. Here are some of the highlights:

* In 1877, a drought and resulting famine in China killed some 10 million people.

* In the 1950s, a severe drought affected parts of Texas, New Mexico, and western Oklahoma, with gigantic walls of dust rolling in from the west.

* In the 1970s and 1980s, a widespread drought devastated the agriculture in Africa's Sahel region, which induced widespread starvation and disrupted entire societies.

* From June to September, 1980, the central and eastern United States tallied $20 billion in drought-related damages and an estimated 13,000 deaths. In the summer of 1988 (with July of that year the hottest month on record in the Northern Hemisphere), the region suffered even greater losses, with $40 billion in damages and 5,000 to 10,000 deaths.

* The summer drought and heat wave of 1993 in the southeast caused $1 billion in damages, with the death toll undetermined.

- During the fall of 1995 through the summer of 1996, severe droughts hit the southern plains, with Texas and Oklahoma the most severely affected. The result: close to $40 billion in estimated damages but, amazingly, no deaths.
- In 1997, northern China suffered from its worst drought in recorded history—the Yellow River, parched by the scorching Sun, had been drained by rampant development—creating losses of more than $175 million and more than 50 deaths.
- In 1998, during an extremely strong El Niño (although not everyone agrees that the storm pattern was the definite cause), drought conditions throughout Micronesia became severe. Even northeastern Brazil had its worst drought ever that year, which wiped out all the crops and livestock in some areas, leading starving farmers to migrate into population centers to survive.

Polar Ice: Another Kind of Desert

Upstate New York was once visited by Ice Age glacial ice sheets, slabs of slow-moving ice that alternately advanced and retreated for more than 100 million years. It plowed the Susquehanna and other rivers into sinuous channels, where they say the ice was over a mile thick. It dammed up and changed the courses of rivers, deepened the Finger Lakes, and pushed sediments, leaving piles of rubble in some places, carving out bedrock in others. It was a cold climate, once occupied by woolly mammoths, "small" mammals (like six-foot-long beavers), and humans, until 10,000 years ago, when the last of the ice sheets retreated. It was another type of desert, one of cold and ice. And the weird weather occurring during the Ice Age would no doubt make our current winters seem like a tropical paradise.

The modern remnants of the Ice Ages include several icy, desertlike regions, in particular the icy polar cap of Antarctica, the area that lies above 75 degrees north latitude (the Arctic), as well as some spots in the highest mountains. These regions exhibit some of the weather phenomena common during the height of the Ice Ages.

The overall climate of the northern and southern polar regions is desertlike, especially Antarctica, at the South Pole. The poles constitute the Earth's coldest, driest, and windiest environment, with average temperatures ranging from −67°F to −76°F (−55°C to −60°C), no more than two inches of snowfall on the average, and winds often exceeding hurricane speeds. The air masses over the Arctic spread over the polar sea ice, while in the Antarctic the air masses are situated over vast, high ice sheets. Because of these polar conditions, strong temperature inversions develop at both poles. Precipitation falls as snow, and though there is very little of it, the continuous cold allows it to accumulate. In the case of Antarctica, the high surface area of the ice sheet (Antarctica covers about 12 percent of the Earth's surface) intensifies the cold, with strong cyclones and blizzard winds quite frequently crossing the ice cap, whose westerly winds are considered the strongest sustained winds on the planet.

• *For more information on ice and snow, try linking to the National Snow and Ice Data Center:* http://www-nsidc.colorado.edu/

• Ice Cycle •

In general, ocean ice forming around the northern and southern poles comes in two forms: *sea ice* and *icebergs*. Sea ice forms during the direct freezing of ocean water. It can also form as *pack ice*, or sea ice that covers the sea surface, breaking up into individual patches called ice floes when the season warms. In contrast, icebergs break off from land ice in a process called calving. The biggest difference between the two types of ice is thickness: While sea ice is usually less than 15 feet (5 meters) thick, icebergs can be hundreds of feet thick.

But not all icebergs are alike. Those forming in the Northern Hemisphere are largely broken from the Greenland Ice Sheet; they are irregular in shape and usually show pointed outlines above the water. In the Southern Hemisphere, icebergs come mainly from the Antarctic ice shelves along the continent's coast, and are commonly tabular in form, with flat tops and steep sides. The ice shelves themselves are actually great plates of floating ice. These shelves are fed by the ice sheet, but they also accumulate new ice from the compaction of fallen snow.

Polar Weather: North and South Pole Chill

Ice, of course, is a given in the colder regions of the North and South Poles, where temperatures are below freezing most of the year. We are all familiar with graupel, hail, and snow, since most of us have lived or heard about such middle-latitude cold-weather terms. But in the cold regions, the cold and winds produce some more bizarre icy features—precipitated,

Pack ice forms in the Great Lakes almost every spring, in the form of broken chunks of melting winter ice. (photo courtesy of the National Oceanic and Atmospheric Administration)

This is a miniature version of an iceberg. The majority of the ice hangs into the water, with only a small amount visible on top of the water. Likewise, icebergs are constantly carved by underwater currents of the oceans; on top, they are carved by icy winds.

so to speak, by strange "icefalls" (instead of rainfall), dry conditions, and uncommon angles of the Sun.

Polar regions are distinguished by their very clear, cold mornings. After the Sun rises, ice prisms called diamond dust fall, creating a sparkling in the sky. You can also observe Sun pillars similar to those seen in the winter months in the temperate latitudes. Later in the day, as the sun remains low on the horizon, vivid Sun dogs are often sighted.

Polar ice comes in a wide and fascinating range of varieties. For example, man-made and natural structures can be covered with feathers of *rime ice*, a milky or opaque deposit of ice formed by the rapid freezing of supercooled water on an exposed surface. Their forms and extent depend on temperature, wind speeds, the size of the water droplets, and turbulence. One simple way to see how rime grows is to examine its domestic equivalent. Just check out small twigs or plants near a waterfall in the winter. The spray of droplets from the falls will freeze on contact, growing from the twig toward the waterfall.

• Even Rougher Polar Weather •

In some of the more frigid regions of the world, blizzards and ice storms can be spectacular. For example, blizzards, which often last for days, are notorious in the southern polar regions, stopping explorers and scientists from conducting experiments. Among the banes of polar travelers are the snow structures called *sanstrugi*, compacted, icy formations resembling sand dunes. And like a sand dune, a sanstrugi moves along as the winds blow, changing its position to fit the whim of the winds and getting in the way of explorers.

Even whiteouts in the polar regions differ from their milder middle-latitude cousins. A special form of whiteout in these regions is called the *ground blizzard*, with visibility reduced to less than 6 feet (1.8 meters). Since these blizzards can persist for days, it's a good idea to make an igloo or pitch a tent for protection, as long as you occasionally check the swirling snow depth so it won't overcome your habitat.

• For more information on Antarctic weather, try linking to:
http://www.compuquill.com/aohtml/aoww0000.html
or the Antarctic Meteorology Research Center:
http://uwamrc.ssec.wisc.edu/amrchrome.html

10

Ocean Weather from the Deep

*P*atricia's father was in the Pacific theater during World War II and spent three long years in the jungles of that ocean's tropical island, right in a main typhoon pathway. Amazingly, he and the other troops endured only two gigantic typhoons. The first came while he was on an almost-empty troop ship, walls of waves crashing on the deck as the ship bobbed up and down like a cork. The second came just after the battle at Guadalcanal. The sky turned to twilight, the winds picked up, and the rains seemed to come from all directions, pelting and stinging the men's faces. Coconut trees fell and bushes were torn from the ground. As the soldiers watched from along a small section of the flat beach, the ocean rose quickly to the tops of the nearby high dunes. Grabbing everything they could, the troops ran for high ground. But the camp flooded, and equipment, tents, and personal belongings were washed out to sea. Trucks and larger machines sank in the mud and sand. No one was hurt or killed—and ironically, the aftermath looked as if there had been a war.

Tropical Storms: Tropical Spawning

The interaction between winds and warm water creates *tropical storm systems*. In the Atlantic, Caribbean, Gulf of Mexico, and eastern Pacific Ocean these systems are called *hurricanes*. In the countries around the Indian Ocean, these storms are called *cyclones*; in the western Pacific and China Sea area, *typhoons* (from the Cantonese *tai-fung*, meaning "great wind"); and around Australia, *willy-willys* (this is also a northern Australian term to describe local dust devils or widespread dust storms).

Tropical storm systems begin with large-scale evaporation and convection of the very warm ocean surface. Great clouds form, eventually producing huge clusters of thunderstorms. *Tropical disturbances* move through the tropics for at least twenty-four hours but show no discernible circular motion. If they produce weak storm systems with winds of less than 38 miles (61 kilometers) per hour, and the winds spin in a circular motion only at the ocean's surface, they are termed *tropical depressions*. These storms develop at least 5 degrees north or south of the equator, where the trade winds of the Northern and Southern Hemispheres meet. From there, if conditions are just right, *tropical storms* form, with the winds definitely spinning and clocked at 39 to 74 miles (63 to 119 kilometers) per hour. And finally, the *tropical cyclones* begin, gathering strength from the ocean's warm water. These are large-scale circular windstorms that travel in the tropics and subtropics, with winds of at least 74 miles (119 kilometers) per hour.

These tropical storm systems don't usually develop beyond

30 degrees north or south of the equator, since there is not enough warm water to fuel them, although rare ones have been encountered over the cooler South Atlantic Ocean or the southeastern Pacific Ocean. They do not form at the equator, either, since the storm systems cannot develop a spin there. In fact, because of this, it is impossible for a tropical storm to cross the equator and survive. If it tries, it merely splits back into a smaller cluster of thunderstorms again.

• For more information on tropical cyclones from
the Hong Kong Observatory, try linking to:
http://www.info.gov.hk/hko/informtc/informtc.htm

• For information on tropical weather watches, try linking to:
http://www.fema.gov/fema/trop.html

• For more information on ocean weather, try linking to:
http://www.oceanweather.com

Atlantic/Pacific Hurricanes: Spinning Waters and Winds

Hurricanes developing over the tropical oceans are cousins of the spinning tornadoes, albeit larger and more complex. After all, hurricanes usually last for hours to days, sometimes coming and going, then returning again. Tornadoes, while destructive, are definitely more short-lived, though the supercells that produce them may last for hours.

Moist air over a warm ocean triggers hurricanes. It sounds simple, but the processes are much more complex: First the humid air flows over the ocean, rising as it heats up. As the water vapor ascends, it condenses, generating not only rain but heat. The warmer upper air creates a pressure area pushing air away from it. In turn, this air, pushed away, cools on the periphery and falls. Rising warmer air replaces it, which creates a low-pressure area near the ocean's surface. The cycle is now in place.

This seemingly complicated routine is similar to what happens when you boil water on your stove. As the water heats at the bottom of the pan, it rises, while the colder water is "pushed" to the bottom in a movement called convection. The water soon boils. As you add more heat to the boiling water and the convection increases, the water boils more furiously. In the storm, this process of convection takes over, feeding the clouds, causing them to grow. Because of the natural motion of the air, the forming hurricane system eventually begins to spin, with smaller convection zones forming as spiral arms. At sustained winds of 74 miles (119 kilometers) per hour, the storm is officially classified as a hurricane.

For the most part, the tropical storms that affect the United States originate just off the west coast of Africa and travel to just outside the Caribbean Sea. This occurs mainly from June to October, when the ocean waters are the warmest. The rotation of the storm signals the possible formation of a hurricane, usually in a region 5 to 25 degrees north of the equator. About a hundred tropical disturbances form each year, but fewer than 10 percent have any potential as hurricane material. And, in an average hurricane year, most of them fizzle out too. Yet if

Hurricane Linda was a major hurricane that formed off the coast of Mexico in the eastern Pacific Ocean in September 1997. This huge system, its eye readily seen in this image, was labeled a Category 5, the highest hurricane level. Fortunately, the hurricane stayed off the coast. (photo courtesy of the National Oceanic and Atmospheric Administration)

they do form, the massive storms can produce sustained winds of up to 150 miles (250 kilometers) per hour, as well as wind gusts at 190 miles (300 kilometers) per hour. They can also drop an average of 2.4 trillion gallons (9 trillion liters) of rain each day. Unfortunately for us, they can last for weeks and cover thousands of square miles.

• *For information on hurricane hunters, try linking to:*
http://www.hurricanehunters.com/welcome.htm
or for the National Hurricane Center (under the Tropical Prediction Center) in the United States:
http://www.nhc.noaa.gov/

• Billion-Dollar U.S. Hurricanes, 1980–1997 •

When a tropical storm's wind speeds reach 39 miles (63 kilometers) per hour, it is assigned a name by the weather service. If it eventually becomes a hurricane, the storm will keep its name. If the hurricane slams into the coast or creates major havoc, its name will be retired—as it was for Hurricanes Bob and Andrew.

Hurricane Name	Date	Area	Costs
Fran	September 1996	North Carolina, Virginia	$5.0 billion; 37 deaths
Opal	October 1995	Southeast U.S.	$3.0 billion; 27 deaths
Marilyn	September 1995	U.S. Virgin Islands	$2.1 billion; 13 deaths
Iniki	September 1992	Kauai, Hawaii	$1.8 billion; 7 deaths
Andrew	August 1992	Florida, Louisiana	$27 billion; 58 deaths
Bob	August 1991	North Carolina, Long Island, New England	$1.5 billion; 18 deaths
Hugo	September 1989	North and South Carolina	$7.1 billion; 57 deaths
Juan	October–November 1985	Louisiana, southeast U.S.	$1.5 billion; 63 deaths
Elena	August–September 1985	Florida to Louisiana	$1.3 billion; 4 deaths
Alicia	August 1983	Texas	$3 billion; 21 deaths

Source: NOAA, National Climate Data Center, Asheville, North Carolina.

Colliding Storms: When Hurricanes Collide

The oceans are huge but sometimes not large enough for two traveling typhoons or hurricanes. Tropical storm Iris approached the Windward Islands on August 23, 1995. At the same time, Hurricane Humberto followed close behind. The merging dance began as Humberto lifted northward over Iris, while the tropical storm slowed down, turning to the south. The tango was short-lived, as the merger weakened the two storms, breaking them apart and sending them on their way. But Iris was not yet finished, for there was still another partner: Eight days later, with winds of 110 miles (177 kilometers) per hour, Iris moved northward toward Bermuda. Tropical storm Karen, a weak storm of only 45 miles (72 kilometers) per hour, cut in behind Iris. The weaker storm relented, absorbed by Iris's strength.

During busy weather seasons, we often see the *Fujiwara effect*, a dance between traveling storms. Studied since the 1920s, the effect is named after Dr. Sakuhei Fujiwara, the chief of the Central Meteorological Bureau in Tokyo after World War I. Observing the motions of vortices (similar to water swirling down a drain), he noted that when two swirling storms came together and rotated around each other, various scenarios could play out, depending on the strength, direction, and combination of spinning movements. We know today that storms will usually dance when they come within 900 miles (1,448 kilometers) of each other. Locked together, both storms actually move about a central point between them, as if they were tied to the same post but were traveling separately around it.

Such a dance can even include smaller unrelated storms. Just before Christmas 1994, a large storm formed off Florida's east coast. Almost like a late-season hurricane, the storm lurched up the coast, creating another such storm in its wake. As the second storm grew, it grabbed its parent, slinging it toward New England. The highest winds in the total storm were clocked at 99 miles (159 kilometers) per hour in Westport Harbor, Massachusetts. Around 130,000 homes lost power in Connecticut, and 5 inches (13 centimeters) of rain fell.

• The Saffir/Simpson Hurricane Scale •

Whether they're merely categorizing storms or trying to justify an evacuation, meteorologists need a way to keep track of damaging hurricanes. Herbert Saffir developed such a scale in the early 1970s. And by 1972, the National Hurricane Center (now under the Tropical Prediction Center) adopted the scale and has used it ever since.

Category 1 Hurricane: winds 74 to 95 miles (119 to 153 kilometers) per hour; minimal damage, limited primarily to shrubbery, trees, foliage, and unanchored mobile homes. Some damage to poorly constructed signs, but no real damage to building structures.

Category 2 Hurricane: winds 96 to 110 miles (154 to 177 kilometers) per hour; moderate damage. Considerable damage to shrubbery and tree foliage, with some trees blown down. Major structural damage to exposed mobile homes and to poorly constructed signs. Some damage to roofing materials, windows, and doors, but not to structural integrity of buildings. (Hurricane Marilyn, which struck the United States Virgin Islands, was a Category 2 hurricane.)

Category 3 Hurricane: winds 111 to 130 miles (178 to 209 kilometers) per hour; extensive damage. Foliage is torn from trees and shrubs, with large trees blown down, as are poorly constructed signs. Some damage to roofing materials and to windows and doors, along with some structural damage to small buildings, residences, and utility buildings. Mobile homes are destroyed. (Hurricane Fran, which struck North Carolina and Virginia, was a Category 3 storm.)

Category 4 Hurricane: winds 131 to 155 miles (210 to 249 kilometers) per hour; extreme damage. Shrubs and trees are blown down, with all signs down. Roofs, windows, and doors are damaged, as are complete roofs on smaller residences. Mobile homes are completely destroyed. (Hurricane Andrew, which struck Florida and Louisiana, was a Category 4 hurricane.)

Category 5 Hurricane: also called super-hurricanes, with winds in these systems greater than 155 miles (249 kilometers) per hour; catastrophic damage. Shrubs and trees are blown down, with all signs down. Roofs collapse or are damaged, and severe window and door damage. Small buildings can be overturned or blown away, and mobile homes are completely destroyed. (Hurricane Linda, which moved through the Pacific Ocean off the coast of southwestern California, was at its peak a Category 5 hurricane. Hurricane Camille, a level 5 hurricane in 1969, was the most lethal storm on record to hit the U.S. mainland.)

• For more information on the Saffir/Simpson Hurricane Scale, try linking to: http://www.nhc.noaa.gov/aboutsshs.html

Storm Surges: Walls of Water

In super-hurricanes, *storm surges* have been known to exceed 20 feet (6 meters). One such hurricane was Camille in 1969: The hurricane had sustained winds of 200 miles (322 kilometers) per hour, and a 27-foot (8.2-meter) storm surge — accompanied by 15-foot (4.6-meter) waves on top of the surge — that smashed into the Mississippi coast at Pass Christian. One story tells of a group of people in a three-story apartment complex who didn't heed the warning. They staged a hurricane party, only to have the waves sweep away everything down to the foundation. Only one woman survived as she floated out a third-floor window on a mattress.

Hurricanes by themselves do not bring death and destruction along a coastline. In some cases, the culprit is a storm surge, in which hurricane winds push water toward the coastlines, surging ashore farther than the normal tides. Meteorologists consider the storm surge one of the most potentially dangerous parts of any tropical cyclone. A combination of hurricane and storm surge killed nearly 6,000 people in Galveston, Texas, in the country's most deadly hurricane in 1900. Cyclones and storm surges also killed 8,000 people in November 1977 and thousands more in 1996, both in the Andhra Pradesh region of India.

Storm surges are appropriately named. A mound of water builds under the center of the hurricane as the low-pressure area draws the water up, like liquid being sucked into a straw. The dome of water builds up as the storm heads toward shore; when the storm reaches the shoreline, the mound comes ashore under the eye of the hurricane. Here, the swells amplify (grow higher) as they approach the relatively shallow continental shelf just off shore, where the friction of the bottom helps to grab the water and pile it up. Not all coastlines have gently sloping continental shelves like the one near Galveston. For example, the rapid drop-off around the island of Jamaica does not allow water to pile up onshore. But for susceptible shorelines, the storms bring huge mounds or walls of water surging inward toward the coastline.

The combined hammering force of the storm surge, the hurricane winds, and the breaking waves can be devastating, like a giant bulldozer sweeping everything in its path. The increase in storm surge power is not linear but exponential: A 20-foot (6-meter) wave has triple the destructive power of a 12-foot (3.66-meter) wave. If you also factor in the actual wave power—a seemingly insignificant 3-foot (1-meter) wave has a pressure of 1,000 pounds per square foot—it's easy to see why surges are so destructive.

One of the worst times to experience a storm surge is at high tide, as the Moon's natural pull adds to the height of the waves. Even worse is *syzygy* (the alignment of the Sun and Moon), which creates the highest tides of all. In syzygy, the dome of water can be more than 50 miles (80 kilometers) wide as it sweeps across the coastline, traveling inland several hundred feet. The Great Miami Hurricane of 1926 occurred at such a time, when the harvest Moon magnified the incoming storm surge to enormous heights.

• Storm Surge Records •

There have been many local records of storm surges along almost all global coastlines, too many to mention here. One of the worst surges on record hit Australia in 1899, during a cyclone. The surge wave was reported to be 40 feet (12 meters) high. And not all surges are caused by strong cyclones. For example, on October 7 and 8, 1996, the Tampa Bay area experienced a substantial storm surge associated with the passage of the "weaker" tropical storm Josephine.

Extratropical Ocean Storms: The "Perfect Storm"

The storm the National Weather Service called "perfect" was an extratropical that turned into a tropical storm. It was also known as the Halloween storm, as it formed on October 28, 1991, and it was the basis for the novel *The Perfect Storm*, by Sebastian Junger, relating the sinking of the swordfishing boat *Andrea Gail*. It wasn't perfect to those who died in the storm, but it was a perfect set of events that resulted in a horrific storm.

An extratropical cyclone developed along a cold front that moved off the northeast coast of the United States. Supported by strong upper air, it deepened. In the meantime, Hurricane Grace, which had formed on October 27 in the Atlantic, made a hairpin turn to the east, responding to the strong wind flows on the developing southern flank of the extratropical low. Grace contributed to the phenomenally bad sea conditions over the western Atlantic. By October 29, a cold front from the extratropical low cut off Grace's low-level circulation, but the

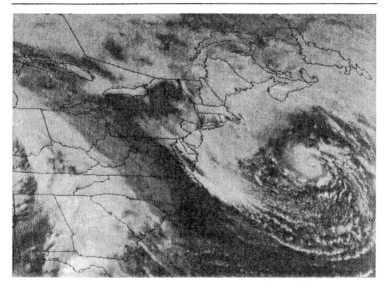

The "perfect storm" was truly one of the most spectacular, according to meteo-rologists: An enormous extratropical low that created havoc along the eastern coast of the United States at the end of October 1991. (photo courtesy of the National Oceanic and Atmospheric Administration, taken by the GOES-7 satellite)

mid- and upper-level moisture was still in place, eventually becoming caught up in the outer part of the extratropical storm center's circulation. This only added to the deepening of the low, with the storm reaching its peak on October 30. By October 31, the storm headed over the warm waters of the Gulf Stream. Although it was never named, it soon became a true hurricane.

The results along the coast were devastating. A high-pressure center extended from the Gulf of Mexico along the Appalachians and into Greenland. Because of the pressure gradient squeeze, the winds increased along the coast. North Carolina's shoreline was lashed with occasional winds of 35 to 45 miles (56 to 72 kilometers) per hour for five days straight.

Hurricane-force winds pounded New England. And high seas were common along the coast from North Carolina to Nova Scotia, with places like New Jersey hit by the highest tides since the Great Atlantic Hurricane of 1944. This anonymous hurricane ultimately destroyed hundreds of millions of dollars in property and killed dozens of people.

Monsoons: Six Months of Rain

In India, *monsoon* season is considered a time of blessing, as the rains break the cycle of oppressive heat across the subcontinent. But it is also a time of tragedy, as the tall Himalayas block the winds, concentrating the destructive rains. In the 1996 season, monsoons killed 2,065 people and destroyed crops covering 6.92 million acres. In 1997, nearly 1,000 people died, and crops covering 3.8 million acres were damaged. And yet, sometimes the monsoon winds don't bring the rains, the consequent dryness killing even more people and crops.

Such are the blessings or curses of the monsoons, weather systems that function almost like large-scale sea and land breezes. The monsoons are caused by the uneven heating of the tropical continents, which causes a dramatic increase in atmospheric pressure over the continents in winter and a decrease in pressure in the summer, relative to the neighboring oceans. The topography of the land creates changes in the weather systems, of course, but not enough to cancel out the overall effects of the entire monsoon season.

India experiences one of the best-known monsoon cycles. It lasts about a year and is equally divided into two seasons. During India's winter monsoon, the Sun is lower on the horizon. The air north of the Tibetan Plateau cools drastically, creating a strong high-pressure system. The resulting winds generally blow southwest over the Himalayan Mountains, then India, and into the oceans. As the winds reach the Indian Ocean, they keep moist air over the oceans. As a result, the weather is clear, dry, and hot on land.

About 80 percent of India's annual rainfall occurs during the other half of the monsoon cycle, the summer monsoon season. From about June to September, the country receives between 150 and 300 inches (381 and 762 centimeters) of rain. The high pressure over the Tibetan Plateau weakens, and low pressure takes over northern India as the summer Sun heats the land. The low-pressure system pulls in the warm, moist air from the Indian Ocean, the winds reversing direction and traveling inland toward the northeast. The end result: long periods of heavy rainfall.

But some years the monsoon rains "stop" just short of the northern end of the continent. In such instances, certain weather systems interfere with the monsoon pattern. Moreover, rain-cooled surfaces slow the rise of air, slowing the monsoon's progress until the land heats up again. If it is too late in the cycle, the rains do not show up at all.

• The "Other" Monsoons •

The word *monsoon* actually derives from the Arabic *mausim*, meaning "season" or "wind shift." This term was first applied to the winds off the Arabian Sea that blow for six months from the northeast and for six months from the southwest. Nevertheless, even Europe has a monsoon, prevailing west-to-northwest winds experienced in the summer months.

In the United States, the "Southwest monsoon" runs northward from Mexico into the southwestern United States and sometimes as far north as Idaho, Montana, and Wyoming during the late summer months. During the winter, the wind flows primarily from the west or northwest; as summer arrives, the winds shift to a southerly or southeasterly direction. The change in winds is due not only to the movement northward of huge upper-air subtropical high-pressure cells, but also to the intense heating of the southwestern deserts.

This monsoon is more accurately called a monsoon thunderstorm, since it is created by convective thunderstorms. In the majority of cases, a clockwise-moving chunk of high pressure from the Gulf of California brings a deep layer of humid air northward, forming scattered clouds during the day. By afternoon, the Sun has heated the ground, and more hot air—thus energy—rises, pushing up water vapor, which condenses into towering thunderstorms. Heavy rains and lightning tend to pop up in the evenings, bringing most of the Southwest's summer rains.

• *For more on the Arizona monsoon, try linking to:*
http://saguaro.la.asu.edu/rcerveny/monsoon.html

El Niño Southern Oscillation: Mixing Up El Niño and La Niña

Contrary to popular belief—and the way the media seem to portray weather patterns to the public—*El Niño* events are *not* a series of catastrophic, flood-producing weather events around the world, although those can and do occur as the *result* of El Niño. In reality, an El Niño event is identified as a rather dramatic increase in ocean surface water temperatures just off the hump of northwestern South America, causing humid air to rise and form storm clouds, coincident with a change in direction of the trade winds across the Pacific Ocean. In non–El Niño years, the winds blow the warm water from South America, across the Pacific, and up against the east coast of Australia; during El Niño years (every two to seven years), the trade winds relax, allowing the heat to flow east, bunching up off the coasts of Ecuador and Peru in a wide swath. Scientists all agree that no one knows why the El Niño begins or how other types of influences—such as global warming or fluctuations in the sun's energy—affect El Niño and, in turn, affect our weather patterns.

Similar to many other weather and oceanic phenomena, El Niño events are classified according to their strength. In 1997–98, the El Niño event was classified as a Type 1 El Niño, with a high sea-surface temperature extending a great distance. What made this El Niño different from the eight other Type 1 events from 1949 to 1993 was its rapid maturity and growth. It was one of the strongest—if not the strongest—El Niño event in the twentieth century. The ocean temperatures averaged almost 11 degrees Fahrenheit (5 degrees Celsius) warmer than usual, measuring 8 degrees Fahrenheit above normal off the South American coast and 5 degrees Fahrenheit above normal off the coast of Baja California.

The results of an El Niño are bizarre. Warm-water fish like marlins enter into the usually cooler waters off the northwestern coast of the United States, the chance of drought increases in Brazil and South Africa, and precipitation and duration of the monsoons of India decrease. Europe can experience wave after wave of stormy weather. In one instance, in early January 1998, tornadoes swept across western Europe and southern England, the latter also experiencing winds at 100 miles (161 kilometers) per hour and violent storms. Australia can receive more rain in some regions: In an El Niño event in early January 1998, 2 feet (0.6 meter) of rain fell on part of Australia in less than twenty-four hours. And more active, intense hurricanes pop up in the strangest places. In November 1997, Hurricane Linda sped toward the coast of California, traveling about 200 miles (322 kilometers) farther north than most hurricanes.

But the El Niño ("Christ child" or "the young boy") is only half of a weather pattern called the *Southern Oscillation,* a cyclical change in the pressure and temperature distribution in the oceans along the equator. As the term implies, the temperature oscillates back and forth over time, though there is no definite period between the peaks of the events—and the characteristics of each event vary greatly. One side of the cycle is called *La Niña* (or "the young child" or "young girl"), during which the surface water of the Pacific Ocean off the coast of South America cools. The other side of the cycle is El Niño, in which the ocean surface waters warm in the same region. Scientists lump the two together, calling the whole phenomenon the *El Niño Southern Oscillation.*

• *For more El Niño sites, try linking to:*
http//www.webpointers.com/elnino.html

• But What About the Atlantic Ocean? •

The Pacific is not the only ocean with a climate beast. The Atlantic Ocean contributes a pattern called the *North Atlantic Oscillation* to the global system. In fact, this oscillation contributes more to the winter weather of Europe and the eastern United States than El Niño. In general, the North Atlantic Oscillation starts as a high-pressure system over the Azores and a low-pressure system over Iceland. When the Icelandic pressure rises and that of the Azores drops, cold Arctic air invades the northeastern United States, pushing weather fronts around. Storms that usually head for England then strike out for Spain, while northern Europe gets extremely cold in the process. Since no one yet knows what drives the weather pattern, no one can truly predict how the oscillation will behave from year to year.

Rogue Waves: From Out of the Depths

Can you imagine being on an ocean vessel when, suddenly, a huge wave hits the ship from out of nowhere? These are *rogue waves*, rare, gigantic walls of water pushing through the oceans. A notorious place for the titanic waves is the Cape of Good Hope off South Africa; other reports include one sighted off the coast of Spain in 1960, when a huge 90-foot (27-meter) wave struck a freighter (to compare, most ocean waves are under 12 feet [3.7 meters]). The ship survived, but strangely enough, the weather was clear throughout the deluge. Just over half a century ago, the *Queen Mary*, pressed into wartime transport service, was nearly capsized off Scotland by what her captain called "one freak mountainous wave." Another rogue wave struck near Daytona Beach, Florida, on July 3, 1992. An 18-foot (5.5-meter) wall of water injured seventy-five people and smashed hundreds of cars parked near the beach. And in 1996, Hurricane Luis generated many waves, including one that the captain of the *Queen Elizabeth II* described as "the white cliffs of Dover coming toward me."

There have been many theoretical explanations for rogue waves. One is underwater landslides, in which large chunks of material fall down an underwater slope and "ripple" the water to create a wave, much like ripples that form when you drop a stone into a pond. Rogue waves may also be caused by the local air pressure effects of a fast-moving line of thunderstorms. The Daytona Beach wave was probably generated in this way, the storm continually adding energy to the moving waves it generated. And many small waves get "in step" with one another after a storm or hurricane, piling up into one gigantic wave. But no matter how the waves are formed, sailors still encounter large "nonnegotiable" mid-ocean waves with no discernible explanation.

Island Wakes: An Island Runs Through It

Ocean clouds and winds do not always generate monsoons, tropical storms, hurricanes, or waterspouts. Occasionally, lighter winds create some of the stranger views of weather as seen from above. One in particular shows up around islands: swirling white cloud vortices that form downwind from islands in the oceans, called *island wakes*. These phenomena were first noticed in the 1960s, when orbiting satellites photographed the oceans. The effect, similar to ship-wake patterns, is best seen from the space shuttle and occasionally from airplanes.

Island wakes come in two forms, or a combination of both. The first is the *von Karman vortex*, named after the German scientist Theodore von Karman, who, in the early 1900s, found such spinning patterns as he watched water moving around stationary objects in his laboratory. *Ship-wake patterns* are the second island wake form. Both these cloud patterns are formed and shaped not by water but by the wind flowing past islands, usually small chunks of land that have some high peaks. Just as river water flowing over and around rocks in a stream creates swirling eddies and ripples, the winds and islands shape the clouds into curling white patterns extending for hundreds of miles.

All island wakes form in the presence of low-level temperature inversions in the atmosphere. These inversions—caps of warm air sitting on top of the cooler air near the ocean's surface—hold back rising air, stabilizing the normal (or prevailing) airflow that the island normally forces upward. The inversion also stops hot air from rising and forming thick, billowy clouds. Instead, the warm air rises just high enough to spread out horizontally, forming long layers of clouds.

In the case of a von Karman vortex, cloud layers meeting a mountain peak need to get around the obstacle somehow. When a mountaintop pokes its way above the cloud layer, the airflow causes the clouds to spill around the mountain slopes, just as water flows around a boulder protruding out of the river. This forms the swirling clouds seen on the island's lee side.

Rows of clouds form downwind on both sides of the island,

An island wake extending from the South Sandwich Islands in the Atlantic Ocean, April 1993. (courtesy of NASA)

shaped by the von Karman vortices. The distance between the rows is about as wide as the island, and the distance between successive swirling clouds in the same row is two or three times as wide as the island. The spiraling cloud patterns remain for long periods of time, sometimes for more than twenty-four hours, extending hundreds of miles from their parent islands. The structure of the swirls finally breaks down as friction from normal air turbulence cuts back on the vortices' rotating energy.

Not all islands form these curling wakes, and much

depends on the prevailing wind speed. If the wind speed is too slow, less than about 16 feet (4.9 meters) per second, the air flow that wraps around the island will lack enough energy to form a swirl; if the wind speed is too fast, more than about 40 feet (12 meters) per second, there is too much energy, and cloud patterns similar to whitewater rapids form. But between these two speeds, vortices readily form.

A ship-wake pattern forms when no part of the island protrudes above the cloud layer. Here, the low-level wind flow encounters merely the highest and lowest spots on the island, creating ripples in the cloud layers. The resulting series of wave patterns are called lee waves, swirling and broken clouds seen for hundreds of miles extending from the island.

Ocean Glitter: All That Glitters

Most of us have seen *glitter* on a fishing trip on the river at sunrise, cruising on a ship at sunset, or even viewing friends' multiple photographs of a sunset on the beach. You would probably notice it on any water-related vacation, just as the Sun is rising, setting, or low on the horizon. Brought on by the multiple images of a low-lying Sun in the water, usually around sunset, or the Moon when it is low on the horizon of an ocean or river, glitter is actually an elongated pattern of light—hundreds of small Suns or Moons reflecting off a stretch of wind-tossed water. Each one of the "glitters" is an instantaneous flash of light reflected from a wave, at just the right position and angle to reach our eyes. There can also be a combination of glitter patterns, as waves in a storm-tossed sea reflect a high Sun.

Glitter patterns in the St. Lawrence River, Thousand Islands, New York, are typical as the Sun sets, showing multiple bright images of the Sun on the wind-tossed waves.

As the Sun or Moon sets lower on the horizon, the glitter pattern tightens up, but not just in one spot. It is seen as a long glittering shaft extending from the horizon just below the Sun or Moon, and almost to our feet. As the Sun or Moon sets on the horizon, the glitter disappears, the light no longer bouncing off wave tops, and the glitter is gone. Glitter can also be seen from an airplane, when the Sun's reflection in the water is seen as an ellipse below the plane.

11

Space Weather and Us

The storm started as a small disturbance on the Sun, a pinprick of a solar disturbance that rapidly turned into an expanding patch of hot gases. Like a polyp split from its parent, the gases stretched out into space, heading outward. The long spike of energetic particles eventually reached the third planet from the Sun, the Earth, as we humans know it. The great geomagnetic storm of March 13, 1989, plunged the entire Canadian Hydro Quebec system and its more than 6 million customers into a blackout, caused by voltage collapses and equipment malfunctions. Though the blackout lasted only about 1.5 minutes, it was enough to give us a taste of our dependence on—and the vulnerability of—our electrical systems. The grid was lucky that time. If the solar storm had been strong enough, the neighboring electrical systems could have experienced a cascade effect that would have rippled all the way down into the United States. Blackouts could have resulted, dropping hundreds of *thousands, perhaps millions, of people into darkness. Such avalanches of solar storms are nothing new and will definitely continue in the future. What needs to be addressed is just what to do about the next huge gaseous tongue from our Sun.*

Solar Wind: A Stellar Performance

The space outside our atmosphere is hardly empty. Particles reign in the realm of space, from minute pieces of planets and satellites to particles from other star systems. Our Sun also puts out a continuous flow of charged particles in all directions, collectively called the *solar wind*. And some of those particles spread out in the direction of Earth.

The major culprits behind the changes in the solar wind are solar flares and coronal mass ejection (CME). The flares are intense, temporary releases of energy, visibly seen as bright ribbons and arcs on the Sun's surface and heard as bursts of noise on shortwave radio. They are the largest explosive events in our solar system, equal to 40 billion Hiroshima-size atomic bombs. Associated with the outer solar atmosphere, and often with solar flares, CMEs are violent and sudden releases of bubbles or tongues of charged particles from the Sun.

The energized material from these phenomena streaks out through space at speeds of a million miles an hour, impacting everything in its path. When this matter reaches Earth, four days after leaving the Sun, this flow of particles distorts our magnetic field, compressing it on the side facing the Sun and spreading it out on the opposite side in a teardrop shape. And like the winds of our atmosphere, the solar wind fluctuates. Stronger gusts of particles can abruptly change the strength and direction of the Earth's surface magnetic field, creating geomagnetic disturbances. At the same time, the stronger solar winds electrify layers of the upper atmosphere (the ionosphere), creating auroras and disrupting communications. The disturbances can also heat the Earth's atmosphere, expanding the air and putting a drag on low-orbiting spacecraft.

• Powerful Solar Cycling •

A typical World War II movie always had the officer barking the order to contact headquarters. The poor radio operator tries frantically to get through with his shortwave, but the signal keeps fading in and out. Eventually, just as the bullets and bombs are flying too close, he contacts HQ. What the moviemakers—and probably the real radio operators in the war—didn't know was that the trouble had to do with the Sun: Solar Cycle 17 was at its peak during the war years.

The solar cycles are actually energetic solar windstorms that appear every eleven years, when the Sun's activity and the number of sunspots (regions of stronger magnetic fields on the Sun's surface) increase. (Actually, it may be a twenty-two-year cycle, the amount of time it takes for the positive and negative charges around the spots to reverse themselves.) The most frustrating aspect of the cycle is that no one knows why the Sun seems to change over a regular interval of time.

The solar cycle was discovered in 1843, but the solar cycle numbers start from the mid-eighteenth century, as some solar data was collected from that time. The latest peak, Solar Cycle 23, is around the years 1999–2001. The most active cycle so far was in

1957, with a peak sunspot number of 201; the estimated sunspot peak for Cycle 23 is 160. Scientists know that the increase in sunspots also equals an increase in solar wind disturbances, hence the concern about the space around our Earth in the early part of the twenty-first century.

Aurora: Stellar Effects

One benefit for scientists working in an observatory is being able to watch what goes on in the nighttime sky while everyone else is warm and snug in bed. The lot of the astronomer, at least in the old days, was to sit out and watch the sky while standing next to a telescope, not sitting at a computer terminal. One of the delights of such personal sky-watching was the *auroral* displays. At first, it seemed as if we were seeing things. A faint glow in the northern sky that might be a cirrus cloud appeared. But the cloud seemed too blue, and it began to waver too quickly even for a fast-moving cloud. Next came the ripples, the curtain of light shimmering and moving in a ghostly fashion. Last, other colors would begin to appear— greens, yellows, and reds—all strengthening our conviction that we were watching an aurora, a rarity at our latitude. It was then that we all promised to make it to the northern regions of the world one day, where almost every night, we could watch the progression of the northern lights.

What causes these colorful glows in the nighttime sky? Like many of the larger planets in the solar system, our Earth possesses a strong magnetic field. The magnetic poles are thought to be a by-product of the Earth's molten, moving mantle and core, both of which create a dynamolike effect. But the actual mechanism is one of the great mysteries of earth science. And although the magnetic poles seem to coincide with the planet's north and south axes, they are not stationary but move by degrees each year around the polar regions.

Further, our Sun releases tiny electrically charged particles of solar wind. These particles travel at high speeds—up to tens of thousands of miles per second—shooting out in every direction, including toward Earth, reaching our planet in about two to four days. Most of the time, the solar winds just pass right by.

But occasionally, these particles react with gases in the Earth's upper atmosphere. As the particles hit nitrogen and oxygen gases in the air at the North or South Poles (between 50 to 600 miles [80 to 1,000 kilometers] in altitude), there is a reaction resulting in beautiful, colorful splashes of light called auroras—the north pole's *aurora borealis* (northern lights) and the south pole's *aurora australis* (southern lights). The flashes range from curtains and waves, to glowing patches and arcs. The yellow-green auroras occur in low-pressure areas when oxygen molecules and electrons collide; red in the aurora is caused as the oxygen molecules and electrons collide in the even lower pressure, high-altitude ranges; and the blues are from electron encounters with atmospheric nitrogen molecules.

The streaming arcs and curtains appear to move and ripple, lasting minutes to hours. And most often above the North Pole, the aurora sometimes meet to form a "boreal crown," or a ring of glowing light centered around the pole, often photographed from the space shuttle.

During strong solar storms, which can happen at any time, the results can be seen much farther south or north from the

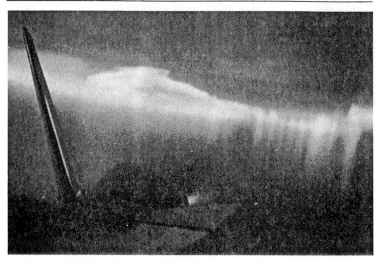

This Southern Hemisphere aurora (aurora australis) was photographed by the space shuttle *Endeavor* in 1994. From a ground vantage point, the aurora would look somewhat similar. If you viewed an aurora, you would see bright blue curtains of light, intermixed with splotches of red, green, and white—all the result of the Sun's particles reacting with Earth's magnetic field. (photo courtesy of NASA)

respective poles. In fact, such a dazzling light show was seen on March 13, 1989, the same night as the power failure at Hydro Quebec, and was visible throughout the entire United States. Overall, many auroral events appear more frequently around the time of the fall and spring equinoxes for reasons that are not clear. And of course, they appear even more frequently during the peak of the eleven-year solar cycle. During the greatest solar cycle peaks, the glowing lights can be seen in the Northern Hemisphere as far south as Athens, Greece, through the entire United States, and into Mexico City, Mexico; in the Southern Hemisphere, auroras can be seen as far north as Brisbane, Australia.

- *For more information on auroras, try the Goddard Space Flight Center:*
http://www.gsfc.nasa.gov

Solar Storms and Space Weather: The Interstellar Weather Report

Contrary to popular belief, the Earth's atmosphere is not really a "blanket" protecting us from the more violent or startling intrusions from space. It is not like a defense grid that holds back large or small asteroids from entering our local space. We know we can't protect ourselves from such large rocks, the evidence being the more than 140 impact holes on our planet's surface.

And the atmosphere is not a barrier against the energetic particles from major solar storms. Space weather truly occurs between our atmosphere and the realms of outer space, reminders that we are part of the solar system and the universe. During the peak of a solar cycle, the Sun can shoot out super storms, energetic solar wind particles that can knock out power facilities, disrupt radio communications, and create havoc for space travelers working on the International Space Station.

In other words, space weather is different from weather in our atmosphere. It's not as predictable; it can create havoc if a storm is strong enough—and *everyone* on the planet is affected by what space sends in our direction.

For some people, the concern over space weather seems to be rather overblown. And some satellite operators and those involved with satellite insurance believe that the threat is not real and newer satellites can better withstand solar events. But our world has changed. Maybe years ago we could get away without thinking about strong solar storms. In the past decade, our reliance on technology has soared, from direct use of our home computers, televisions, and sensitive electronic devices, to the indirect electronics used by our banks or at our jobs.

Scientists now recognize a link between solar activity, geomagnetic disturbances, and electronic disruptions to systems such as power grids, satellites, communications, and defense systems. For example, the super storm that knocked out the Hydro Quebec power facility cut off electricity and heat for about 6 million people for a short time as transformers blew, and garage doors throughout North America automatically opened and closed for hours afterward. PanAmSat's *Galaxy IV* temporarily went on the blink in mid-May 1998 from an energetic solar storm, interrupting paging, TV feeds, and other services—and this was not even happening at a solar cycle peak.

In order to warn against such storms, the National Space Weather Program (NSWP) was developed, gathering information on the major players that can endanger human life and health: solar wind, the magnetosphere, and the upper atmosphere. Although we don't have the Space Weather Channel yet, some scientists and policymakers are already preparing for that day.

The National Oceanic and Atmospheric Administration's Space Environment Center in Boulder, Colorado, is playing such a part in space forecasting. In the near future, you'll see space weather maps in the National Weather Service format. One day, you may wake up to not only the atmospheric weather report but also the forecast of a major solar storm heading

Earth's way. The notices will eventually become a regular feature in weather forecasts in the media, providing the public with explanations for different phenomena caused by space weather. The reports will explain how the storm might affect them, and why such things as your radio, cellular phone, or global positioning systems may be out of whack that day.

And if the skies in Missouri or Kansas turn a bright red one night, the reports will explain to a curious—or nervous—public that the sky was not on fire, nor was there a nuclear explosion, escaping chemicals, or an alien invasion taking place. Rather, it was more likely the bright and shimmering aurora borealis, or northern lights, the display reaching farther south than usual.

• *For more information about space weather, try linking to the Space Studies Board's links:*
http://www.nas.edu/ssb/links.html
or try the Lund Space Weather Center in Sweden:
http://www.astro.lu.se/~henrik/
or the NOAA web site:
http://www.sec.noaa.gov

• How to Watch a Solar Storm •

Watching solar storms has never been easy from Earth. One of the easiest ways to detect the outflow of gases from the Sun is via spacecraft that orbit the Sun or the Earth. For example, NASA's Polar Ultraviolet Imager orbiting the Earth has shown how solar activity that causes auroral displays directly affects the Earth's outer atmosphere. In September 1998, the craft imaged oxygen and other gases being blasted into space as a solar storm pumped about 200 gigawatts of energy into the atmosphere. (A normal aurora produces only a few gigawatts.) Trapped in the wake of the solar wind, most of the gases eventually returned to the atmosphere.

Other Earth-orbiting spacecraft are on solar alert too. During a January 1997 storm, eleven satellites in a variety of orbits contributed space weather data. These and other craft measure "relativistic electron" events—indications that a patch of energetic particles from the Sun are interacting with Earth's upper atmosphere—providing data that can be used on space weather maps.

The Damages: What Solar Storms Could Do

What are some of the effects of massive solar storms? We already know about the effects the storms had on the Canadian grid, but there are other outcomes of such space storms. Here are a few:

Satellites A giant solar storm can cause the Earth's atmosphere to expand, dragging space satellites into lower orbits, like an invisible hand tugging at the craft, often affecting its orientation. The *Skylab* spacecraft was a perfect example when it re-entered the Earth's atmosphere prematurely. Greater solar activity than predicted caused the atmosphere to expand, slowing down the craft and dragging it into a lower orbit—too low to push back into a higher orbit. The great geomagnetic storm of March 1989 almost fried four Navy navigation satellites, leaving them out of commission for a week. The charged particles from such storms can often build up charges on the craft, interfering with the sensitive electronics onboard, a major concern as more spacecraft are launched to keep our technology going.

Pipelines Electric currents induced in the Earth during a magnetic storm can corrode buried conductors, such as pipelines—this despite protective special coatings or even corrosion-inhibiting electric voltage applied along the pipeline.

Communications Solar radio bursts can interfere with VHF and UHF signals in the sunlit hemisphere of the planet; the solar ultraviolet flux during the solar cycle is in the range of frequencies available to HF radio communications, and X rays from solar flares can produce shortwave fadeouts. Television and commercial radio stations are little affected, but ground-to-air, ship-to-shore, Voice of America–type programs, and amateur radio are frequently disrupted. Some military detection or early warning systems can also be affected.

Airplanes Passengers on high-flying aircraft could receive a little extra radiation dose, but they are usually not in flight long enough to make the extra radiation a problem.

Manned spacecraft The particles released by a major burst of solar activity are potentially lethal to the crews of space vehicles. Astronauts on the International Space Station, as well as participants in future interplanetary missions, would be directly affected by an increase in radiation.

Magnetic field If a cloud of solar material explodes on the Sun, and if the Earth is in the pathway of the particles, our planet's magnetosphere will feel the effects one to four days later. The result is a geomagnetic storm, causing the Earth's magnetic field to fluctuate wildly.

• The Sun-Earth Climate Connection? •

The aurora and geomagnetic storms are not the only sky events attributed to our Sun that affect Earth. There is also evidence that our planet's climate may be extremely responsive to changes in solar activity. Between the years 1450 and 1850, European historians recorded bitterly cold winters, cooler summers, failed harvests, and violent storms. In particular, the period between 1640 and 1710, during which the Sun had very few sunspots, corresponded to a mini-ice age in Europe called the Maunder Minimum. While the Sun's energy output fell by less than 1 percent, a small number, it was still enough to affect the continent's overall weather. The English Channel froze solid, glaciers extended southward, and many people froze or starved from the cold. Even with all the evidence, the correlation between the absence of sunspots and our climate is still highly debated.

Not that we have to worry. In spite of all the science fiction stories, not all stars in the universe end up exploding in a nova or supernova at the end of their lives. Scientists believe that although our Sun fluctuates in luminosity and energy over time, it still has a long way to go before it grows dark—about 5 billion to 6 billion years.

Solar-System Systems: Weather on Other Planets

One of the first times we saw a dust devil, a swirling column of dust and wind, was not in an Earth desert but in images taken by the *Viking* spacecraft on one of the largest deserts in the solar system, the arid and dusty planet Mars. It was the first time scientists viewed any other type of weather on another planet—up close and personal.

These tall, dust-filled winds form under nearly the same conditions as on Earth, in an arid area with plenty of dust and sand. In the case of Mars, the dust devils most often form during the warmer summer season, when the major global windstorms begin; on Earth, strong local winds cause the dust to be lifted up, swirling from a combination of air mass rotation and strong updrafts. We'd be glad to take the dust devils on Earth over the ones on Mars. The dust devils on Earth don't usually cause too much damage, but the red planet's vortices are more formidable, rising miles high.

What about the other planets of the solar system? Do they have weather systems similar to Earth? Not quite: Earth is a unique planet. But there are some mind-boggling comparisons and contrasts:

Mercury We can eliminate Mercury from the list. The hot Sun and the planet's small size (lower gravity) caused the planet to lose its atmosphere early in the history of the solar system.

Venus Cloudy Venus has such a thick atmosphere that it would crush you if you stood on the surface. As for weather, the planet has one of the most interesting "rainstorms" in the solar system: It rains sulfuric acid, drops that fall from the clouds and evaporate before they reach the ground.

Mars Fog and frost persist in the thin atmosphere of the red planet. Its atmosphere is mostly carbon dioxide with traces of water vapor and other constituents. It has high humidity because of the extremely low temperatures, and thus, a much lower dew point. Frost forms in low-lying areas during the extremely cold nights; in the early morning, the rays of the Sun vaporize the ground frost, and fog forms over the rocky ground. But unlike on Earth, the fog doesn't disrupt the morning rush-hour traffic.

Jupiter The Great Red Spot (GRS) on Jupiter looks like its name: a bright red blob easily seen from Earth with a small telescope. The oval spot has probably been there for over 300 years and is a huge, spinning storm system that just won't go away. There are other storms that grow, shrink, or dissipate over the years in the upper atmosphere of the planet, but none are as vivid as the GRS.

Saturn The planet is famous for its rings; in terms of weather, it has one of the fastest atmospheres in the solar system. In fact, the winds move more than 1,000 miles (1,609 kilometers) per hour, speeds that would leave nothing standing on Earth. Saturn also occasionally "burps." This term is commonly used by planetary scientists to describe a periodic storm system that seems to pop up every thirty years or so around the equator of the planet, then gradually fades away.

Uranus If you like variation in weather, Uranus is not the place to go. Spacecraft flying by the planet have shown very little detail, mostly just a blank, bright, light blue marble spinning around the Sun.

Neptune Neptune has one of the most dynamic weather systems, filled with scooters, dark spots, and light spots, all of which travel quickly around the planet. Similar to the Great Red Spot on Jupiter, these swirling blemishes are fast-moving storms that form in an atmosphere filled with hydrogen, helium, methane, and ammonia.

Pluto Because of the difficulty in imaging such a small and distant planet, scientists merely hypothesize about an atmosphere around Pluto. They do agree that Pluto is a cold and miserable place, surrounded by a tenuous envelope of pure methane, with surface pressures some 100,000 times less than that at Earth's sea level. As for climate, you'll need more than mittens and a muffler to live on this planet, with warmest temperatures estimated at −350°F (−212°C).

• For more information on Mars and the other planets, try linking to:

http://www.nasa.gov/

• Does the Moon Affect Our Weather? •

According to some studies, the Moon does appear to have some impact on our weather. Statistical analysis shows that the Moon's phases are somewhat correlated to certain weather conditions, including the number of thunderstorms, air pressure changes, and cloudiness. There really is no explanation, but scientists at the University of Arizona suggest that the temperatures in the Earth's lower atmosphere increase by about 0.02 degrees Celsius during certain lunar phases. If this is true, it may explain at least a part of these weather events.

Epilogue

If you look hard enough, you can count on observing some very strange weather, no matter where you live or travel. The best way is, of course, to become aware of your surroundings. Look up when you get out of work, and you might see a sun dog on either side of the setting sun. Keep track of your local pressure with a barometer. Always request a window seat forward of the wing on a plane, so you can see above and below the horizon. On more northern night flights in the Northern Hemisphere, turn the light off and look for the aurora borealis (and in the Southern Hemisphere, the aurora australis). Request a window seat on a train to observe the back end of (usually as you travel east) or approaching (usually as you travel west) warm and cold fronts. Take pictures of your favorite weather sights.

In other words, keep your eyes open, and nature will reward you with some fantastic shows. All it takes is a simple change—in a cloud, a swirl of the wind, or lowering of the sun. We know there are weather sights out there that no one else has described or even thought about. So keep a journal of your own strange skies, and let us know what you find.

Links to Weather

The following are only some of the extra links to general
weather sites on the Internet:

1. *World Weather link*:
   http://www.geocities.com/SiliconValley/3452/
   weather.html
2. *Weather links around the world*:
   http://sweb.srmc.com/srmc/weather.html
3. *Golden Triangle page*:
   http://www.ih2000.net/ira/bmt-wth.htm
4. *World Meteorological Organization*:
   http://www.wmo.ch/
5. *National Climate Data Center*:
   http://www.ncdc.noaa.gov/
6. *National Oceanic and Atmospheric Administration*:
   http://www.noaa.gov/
7. *International Weather Satellite Imagery Center*:
   http://www.fas.harvard.edu/~dbaron/sat/
8. *NASA and climate*:
   http://climate.gsfc.nasa.gov/
9. *National Weather Service*:
   http://www.nws.noaa.gov/
10. *National Center for Atmospheric Research*:
    http://www.ncar.ucar.edu/

Note: There are also smaller observatories and institutes, such as the Catalina Island Conservancy Intranet that gathers information on the California island's weather and sundry other weather sites at:
   http://www.catalinas.net/seer/menuclim.ht

Index

Printed in the United States
46675LVS00002B/51